Mathematics for Machine Learning: A Deep Dive into Algorithms

WRITTEN BY NIBEDITA SAHU

While every precaution has been taken in the preparation of this book, the publisher assumes no responsibility for errors or omissions, or for damages resulting from the use of the information contained herein.

Mathematics for Machine Learning: A Deep Dive into Algorithms

by Nibedita Sahu

Copyright © 2023 Nibedita Sahu. All rights reserved.

Written by nibedita sahu

[3]

Preface

Welcome to "Mathematics for Machine Learning: A Deep Dive into Algorithms." In an era defined by data-driven insights and intelligent systems, the marriage of mathematics and machine learning has become an essential cornerstone of innovation. This book aims to guide you through the intricate landscape of mathematical concepts that form the bedrock of modern machine learning techniques.

As technology continues to evolve, understanding the mathematical foundations behind machine learning algorithms has become not just a skill, but a necessity. This book is tailored to students, practitioners, and enthusiasts who seek a comprehensive understanding of how mathematical principles shape the realm of machine learning. Whether you're a seasoned professional or a curious newcomer, this book will provide you with the tools to navigate the complex terrain of mathematical concepts that power machine learning algorithms.

The Journey Ahead

In the pages that follow, you will embark on a journey that explores linear algebra, calculus, probability, statistics, and information theory—all tailored to the unique demands of machine learning. Through clear explanations, illustrative examples, and practical applications, you will build a robust foundation in mathematics that empowers you to tackle real-world challenges in the field of artificial intelligence.

Each chapter is designed to unravel the mysteries of mathematical intricacies while maintaining a direct connection to their application in machine learning. From regression to neural networks, from optimization to reinforcement learning, this book offers a comprehensive roadmap to understanding both the "how" and the "why" of mathematical operations in the context of machine learning.

Getting the Most from This Book

To make the most of your reading experience, take advantage of the exercises provided at the end of each chapter. These exercises serve as opportunities to apply and reinforce the concepts discussed in the text. For those seeking additional clarity, the solutions to selected exercises can be found in the appendix itself.

Remember, this book is not intended to be a mere reference—it's a guide that encourages active learning.

Engage with the material, experiment with code snippets, and embrace the challenge of grappling with complex mathematical concepts. With dedication and persistence, you'll find yourself equipped with the knowledge and skills needed to excel in the world of machine learning.

Let's Begin

Now, with curiosity as your compass and this book as your guide, let's embark on a captivating exploration of mathematics in the realm of machine learning. As you turn the pages and immerse yourself in the topics ahead, remember that every concept mastered brings you one step closer to unlocking the full potential of machine learning algorithms.

Your journey starts here.

Sincerely,

Nibedita Sahu

[6]

[7]

NIBEDITA SAHU

Table of Contents

Introduction [19]

- The Role of Mathematics in Machine Learning
- Prerequisites for the Book
- How to Use This Book Effectively

Chapter 1: Foundations of Linear Algebra [31]

1.1 Vectors and Matrices [32]

1.2 Matrix Operations [36]

1.3 Vector Spaces and Linear Transformations [40]

1.4 Eigenvalues and Eigenvectors [45]

Chapter 2: Multivariable Calculus [50]

2.1 Partial Derivatives [51]

2.2 Gradients and Jacobian Matrices [55]

2.3 Chain Rule and Higher-Order Derivatives [59]

2.4 Optimization Techniques [63]

Chapter 3: Probability and Statistics [68]

3.1 Basic Probability Concepts [70]

3.2 Random Variables and Probability Distributions [74]

3.3 Expectation, Variance, and Covariance [78]

3.4 Maximum Likelihood Estimation [82]

Chapter 4: Information Theory [87]

4.1 Entropy and Information Gain [88]

4.2 Kullback-Leibler Divergence [92]

4.3 Mutual Information and Applications [96]

Chapter 5: Linear Regression [101]

5.1 Simple Linear Regression [102]

5.2 Multiple Linear Regression [106]

5.3 Least Squares Estimation [110]

5.4 Regularization Techniques [114]

Chapter 6: Classification [119]

6.1 Logistic Regression [120]

6.2 Softmax Regression [124]

6.3 Binary vs. Multiclass Classification [128]

6.4 Evaluation Metrics [132]

Chapter 7: Support Vector Machines [138]

7.1 Linear SVM [139]

7.2 Non-linear SVM [143]

7.3 Kernel Trick [147]

7.4 Margin and Slack Variables [151]

Chapter 8: Neural Networks and Deep Learning Basics [156]

8.1 Perceptrons and Activation Functions [157]

8.2 Feedforward Neural Networks [161]

8.3 Backpropagation Algorithm [166]

8.4 Training Neural Networks [171]

Chapter 9: Convolutional Neural Networks (CNNs) [177]

9.1 Image Representation [178]

9.2 Convolution and Pooling Layers [182]

9.3 CNN Architectures [187]

9.4 Transfer Learning with CNNs [192]

Chapter 10: Recurrent Neural Networks (RNNs) [198]

10.1 Sequential Data [199]

10.2 RNN Architecture [204]

10.3 Long Short-Term Memory (LSTM) Networks [208]

10.4 Applications in NLP [213]

Chapter 11: Unsupervised Learning: Clustering and Dimensionality Reduction [219]

    11.1 K-Means Clustering [220]

    11.2 Hierarchical Clustering [225]

    11.3 Principal Component Analysis (PCA) [230]

    11.4 t-Distributed Stochastic Neighbor Embedding (t-SNE) [235]

Chapter 12: Regularization and Regularized Regression [240]

    12.1 Ridge Regression [241]

    12.2 Lasso Regression [244]

    12.3 Elastic Net [247]

    12.4 Choosing Regularization Parameters [250]

Chapter 13: Decision Trees and Ensemble Learning [255]

    13.1 Decision Tree Construction [256]

    13.2 Random Forests [259]

    13.3 Gradient Boosting [262]

    13.4 XGBoost and LightGBM [265]

Chapter 14: Neural Network Architectures [270]

    14.1 Autoencoders [271]

14.2 Generative Adversarial Networks (GANs) [275]

14.3 Transformers [278]

14.4 Applications in Generation and NLP [281]

Chapter 15: Future Trends in Machine Learning [286]

15.1 Explainable AI [287]

15.2 Federated Learning [291]

15.3 Quantum Machine Learning [295]

15.4 Ethical Considerations [299]

Appendix : Mathematical Notation, Concepts, examples, exercises and solutions [304]

- Common Symbols and Notations
- Review of Key Mathematical Concepts
- Exercises and Solution (Brief explanations)

Book Summary

"Mathematics for Machine Learning: A Deep Dive into Algorithms" is your comprehensive guide to understanding the intricate relationship between mathematics and the world of machine learning. In an age defined by data-driven insights and intelligent systems, this book equips both beginners and experienced practitioners with the essential mathematical toolkit to excel in the field of artificial intelligence.

Key Features

Foundational Concepts: Delve into the core principles of linear algebra, multivariable calculus, probability, statistics, and information theory. Explore vectors, matrices, derivatives, expectations, and entropy, all tailored to their applications in machine learning.

Real-world Applications: Connect mathematical abstractions to tangible machine learning techniques. From linear regression and classification to neural networks and reinforcement learning, each chapter illustrates how mathematics drives the algorithms behind data analysis, pattern recognition, and decision making.

Practical Examples: Grasp mathematical concepts through practical examples and case studies. Understand how mathematical operations are translated into code and applied to real-world datasets, making abstract ideas come to life.

Exercises and Solutions: Put your knowledge to the test with exercises at the end of each chapter. Strengthen your understanding through hands-on problem solving, and refer to the solutions in the appendix to reinforce your learning.

Ethical Considerations: Navigate the ethical landscape of machine learning. Learn how mathematical algorithms can have societal implications and gain insights into addressing bias, fairness, and transparency in AI systems.

Future Directions: Explore emerging trends in machine learning, from explainable AI to quantum machine learning. Understand the potential impact of these trends and how they intersect with the mathematical foundations laid out in the book.

Your Journey Begins Here

Whether you're a student, a developer, a data scientist, or a dedicated enthusiast, "Mathematics for Machine Learning" offers a roadmap for your journey through the symbiotic realms of mathematics and AI. With clear explanations, insightful examples, and a strong emphasis on application, this book empowers you to not only grasp the intricate mathematics behind machine learning but also apply it to build intelligent systems that transform data into meaningful insights.

Prepare to unlock the full potential of machine learning algorithms through a deep dive into the mathematics that powers them. Welcome to a world where equations drive innovation, and intelligence meets insight.

MATHEMATICS FOR MACHINE LEARNING: A DEEP DIVE INTO ALGORITHMS

Nibedita Sahu

[18]

Introduction

The Role of Mathematics in Machine Learning

In the realm of modern technology and artificial intelligence, machine learning has emerged as a transformative force, enabling computers to learn from data and make intelligent decisions without being explicitly programmed. At the heart of this remarkable advancement lies the intricate interplay between machine learning and mathematics. Mathematics serves as the foundational framework that empowers machine learning algorithms to comprehend patterns, derive insights, and make accurate predictions from vast and complex datasets.

Mathematics provides the essential language for expressing the principles and algorithms that underpin machine learning. From linear algebra to calculus, and from probability theory to optimization, mathematical concepts form the bedrock upon which machine learning methodologies are built. Here, we delve into the multifaceted role that mathematics plays in shaping the landscape of machine learning, uncovering the key insights that make this symbiotic relationship so crucial.

- ### *Linear Algebra:*

Linear algebra is the cornerstone of many machine learning algorithms. Vectors and matrices allow data to be represented and manipulated efficiently. Concepts like eigenvectors and eigenvalues find applications in dimensionality reduction techniques like Principal Component Analysis (PCA) and in understanding the behavior of neural networks. Matrix factorization techniques are used in recommendation systems, a prime example being collaborative filtering.

- ### *Calculus:*

Calculus provides the tools for understanding rates of change and optimization. Gradient descent, a fundamental optimization technique, uses derivatives to iteratively refine model parameters and minimize error. This is at the heart of training neural networks and other learning algorithms. Integral calculus finds its place in probability distributions, crucial for generative models and probabilistic reasoning.

- ### *Probability and Statistics:*

Probability theory is the linchpin of uncertainty handling in machine learning. Bayes' theorem forms the basis of Bayesian

machine learning, a paradigm that incorporates prior knowledge to update beliefs as new data arrives. Statistical concepts like mean, variance, and standard deviation are used for data summarization, and distributions such as Gaussian and Poisson distributions model various data types.

- **_Optimization:_**

Optimization algorithms seek to find the best solutions among a set of possible choices. Gradient descent, as mentioned earlier, is a classic optimization technique, while more advanced methods like stochastic gradient descent (SGD) and Adam optimization enhance the training of deep neural networks. Convex optimization plays a role in support vector machines and linear regression.

- **_Information Theory:_**

Information theory quantifies the amount of information in data and the efficiency of data compression. Concepts like entropy, mutual information, and the Kullback-Leibler divergence are used in various applications, such as feature selection and evaluation metrics for clustering and classification.

- ***Geometry and Topology:***

Geometric insights aid in understanding the relationships between data points. Non-linear dimensionality reduction techniques like t-SNE (t-Distributed Stochastic Neighbor Embedding) leverage manifold learning to preserve local structures in high-dimensional data. Topological data analysis explores the shape of data through concepts like persistent homology, enabling the discovery of hidden patterns.

- ***Graph Theory:***

Graphs provide a way to model relationships between entities. Graph algorithms are used in recommendation systems, social network analysis, and natural language processing tasks like parsing syntax trees. Graph neural networks extend deep learning to graph-structured data.

- ***Numerical Methods:***

Many machine learning algorithms involve solving numerical problems. Techniques from numerical analysis ensure stable

[23]

and accurate computations, critical for reliable model training and inference.

In conclusion, the role of mathematics in machine learning is undeniably profound. Mathematics furnishes machine learning practitioners with a set of tools that enable them to extract meaningful information from raw data, build predictive models, and uncover hidden insights. The synergy between mathematics and machine learning is not just about applying mathematical concepts, but about leveraging the inherent beauty and logic of mathematics to construct algorithms that mimic cognitive processes, leading to machines that can learn, adapt, and evolve. As machine learning continues to evolve, its symbiotic relationship with mathematics will remain a cornerstone of innovation in this dynamic field.

Prerequisites for the Book

Before you embark on your journey through "Mathematics for Machine Learning: A Deep Dive into Algorithms," it's important to have a clear understanding of the background knowledge and skills that will best prepare you for the content of this book. While no prior expertise in machine learning is assumed, a foundational understanding of mathematics and a basic familiarity with programming concepts will greatly enhance your experience.

Mathematical Background

A solid grasp of foundational mathematical concepts forms the backbone of this book. Prior exposure to topics such as algebra, calculus, linear algebra, probability, and statistics will be beneficial. While we will explain these concepts in the context of machine learning, having a foundation will help you dive into the more advanced applications and discussions.

Curiosity and Dedication

Above all, an open and curious mind combined with dedication to learning will be your greatest assets. Machine learning is a dynamic and evolving field, and this book aims to equip you with the tools to adapt and explore beyond its pages. Be prepared to engage with challenging concepts, work through exercises, and experiment with code to solidify your understanding.

Remember, this book is designed to guide you from the fundamentals to advanced applications, providing you with the necessary mathematical and practical insights along the way. With the right mindset and a willingness to learn, you are well-prepared to embark on this enlightening journey through the mathematics that underpin machine learning.

How to Use This Book Effectively

"Mathematics for Machine Learning: A Deep Dive into Algorithms" is designed to be a comprehensive resource that takes you on a journey through the mathematical foundations of machine learning. To make the most of this book and maximize your learning experience, consider the following tips:

1. Sequential Progression: The book is structured in a logical progression, starting from foundational mathematical concepts and gradually moving to more advanced topics. It's recommended to read the chapters in order, as later chapters may build upon concepts introduced earlier.

2. Active Engagement: Engage actively with the material. Take your time to read through explanations and examples, and make sure you understand each concept before moving on to the next. Don't hesitate to take notes, highlight key points, and jot down questions that arise.

3. Exercises and Practice: Each chapter includes exercises that allow you to apply the concepts you've learned. These exercises are designed to reinforce your understanding and help you gain hands-on experience. Attempt the exercises and refer to the solutions in the appendix for verification.

4. Connections to Real-World Applications: As you study each mathematical concept, pay attention to how it relates to real-world machine learning techniques. Understanding these connections will help you bridge the gap between theory and practice.

5. Reflect and Review: Periodically take time to reflect on what you've learned. Review key concepts, revisit examples, and consider how different topics fit together. This reinforcement can help solidify your understanding.

6. Discussion and Exploration: Engage in discussions with peers, online communities, or mentors. Exploring different perspectives can enhance your learning and provide additional insights into complex topics.

7. Appendix and Additional Resources: Use the appendix as a quick reference for mathematical notation and concepts.

Additionally, consider exploring supplementary resources, such as online tutorials, courses, and academic papers, to delve deeper into specific areas of interest.

8. Practice Patience: Some concepts may initially seem challenging, but don't be discouraged. Mastery takes time and practice. Approach each topic with patience and persistence, and don't hesitate to revisit sections as needed.

9. Apply Beyond the Book: The ultimate goal is to apply what you've learned beyond the book. Consider implementing machine learning algorithms, experimenting with datasets, and contributing to projects that align with your interests.

Final Thoughts

"Mathematics for Machine Learning" is a resource to empower you with the knowledge and skills needed to navigate the intricate landscape of mathematical concepts in the world of machine learning. By actively engaging with the material, seeking to understand the "why" behind each concept, and translating theory into practice, you'll not only grasp the mathematics but also unlock the potential to create innovative solutions through intelligent algorithms.

[29]

Remember, this book is a guide, and your journey doesn't end with its final page. The knowledge you gain here will be a stepping stone to further exploration and discovery in the dynamic realm of machine learning.

Happy learning!

[30]

Chapter 1: Foundations of Linear Algebra

1.1 Vectors and Matrices

1.2 Matrix Operations

1.3 Vector Spaces and Linear Transformations

1.4 Eigenvalues and Eigenvectors

1.1. Vectors and Matrices

Vectors and matrices are fundamental mathematical structures that serve as the building blocks for representing and manipulating data in machine learning. They provide a concise and efficient way to organize, transform, and analyze complex datasets, enabling machine learning algorithms to make sense of raw information and extract meaningful patterns. Here's how vectors and matrices play a pivotal role in various aspects of machine learning:

Data Representation: In machine learning, data often comes in the form of attributes or features that describe the characteristics of the objects being studied. Vectors are used to represent these attributes in a structured manner. Each dimension of the vector corresponds to a specific feature, allowing the algorithm to work with multiple attributes simultaneously. For example, in image classification, each pixel's color intensity could be represented as a dimension in a vector.

Feature Engineering: Feature engineering involves selecting, transforming, and creating new features to enhance the performance of machine learning models. This process often involves operations on vectors, such as normalization (scaling values to a standard range), which helps in improving the convergence of optimization algorithms and ensuring that no feature dominates others.

Matrix Representations: Matrices extend the concept of vectors to two or more dimensions. They are particularly useful for representing relationships between multiple variables. For instance, in a dataset where each row corresponds to an individual and each column corresponds to a different attribute (e.g., age, income, education level), the entire dataset can be represented as a matrix. This matrix-based representation simplifies operations such as matrix multiplication and factorization.

Linear Transformations: Vectors and matrices are central to linear transformations, which are essential for various machine learning tasks. Transformations like scaling, rotation, and translation can be represented using matrices. For instance, in image processing, matrices are used to perform transformations like resizing and rotating images.

Feature Space and Decision Boundaries: In machine learning algorithms like support vector machines (SVM) and neural networks, decision boundaries are defined in a higher-dimensional feature space. These decision boundaries are determined by the relationships between vectors in the input space, transformed using weight matrices and activation functions.

Eigenvalues and Eigenvectors: Eigenvalues and eigenvectors are key concepts from linear algebra that have applications in dimensionality reduction, principal component analysis (PCA), and eigenvalue decomposition. PCA, for example, involves finding the eigenvectors of the covariance matrix of data, allowing for efficient representation and compression of information.

Matrix Factorization: Matrix factorization techniques, such as singular value decomposition (SVD) and non-negative matrix factorization (NMF), are used in various contexts, including collaborative filtering for recommendation systems and topic modeling in natural language processing. These techniques decompose a matrix into smaller matrices that capture latent patterns and relationships in the data.

In essence, vectors and matrices provide a unified framework for organizing and manipulating data in machine learning. They enable algorithms to perform complex operations efficiently, transforming raw data into a format that can be processed to extract valuable insights. Without the foundational concepts of vectors and matrices, the development and application of many machine learning techniques, from linear regression to deep learning, would not be possible. These mathematical constructs truly serve as the backbone of modern data-driven algorithms, bridging the gap between raw information and intelligent decision-making.

1.2 Matrix Operations

Matrix operations are the essential tools that empower machine learning algorithms to analyze, process, and extract valuable insights from data. These operations involve manipulating matrices and vectors through various mathematical operations, such as addition, multiplication, inversion, and decomposition. They underpin a wide range of machine learning techniques and enable algorithms to perform complex computations efficiently. Here's how matrix operations play a critical role in machine learning:

Linear Regression: In linear regression, matrix operations are used to calculate the optimal coefficients that minimize the error between predicted values and actual data points. The least squares method involves solving a system of linear equations, which can be efficiently represented and solved using matrix notation.

Matrix Multiplication: Matrix multiplication is a core operation in machine learning. It enables the transformation of data by applying a set of weights or coefficients to each input feature. This operation is foundational in neural networks, where layers of neurons perform matrix multiplications followed by activation functions to model complex relationships within the data.

Convolutional Neural Networks (CNNs): CNNs, widely used in image and video analysis, leverage convolution operations to extract features from images. Convolutions involve element-wise multiplication of filter matrices with overlapping sections of the input image, allowing the network to detect various patterns and structures.

Dimensionality Reduction: Techniques like Principal Component Analysis (PCA) and Singular Value Decomposition (SVD) involve matrix factorization, which breaks down a matrix into a product of smaller matrices. These operations are crucial for reducing the dimensionality of data while retaining important information, leading to improved efficiency and interpretability.

Eigenvalue Decomposition: Eigenvalue decomposition is a matrix operation that breaks down a matrix into its eigenvalues and eigenvectors. It's used in various applications, such as spectral clustering and solving systems of differential equations, providing insights into the behavior of dynamic systems.

Kernel Methods: Kernel methods involve computing similarity measures between data points in a transformed space. These measures are often derived through matrix operations, allowing algorithms to capture nonlinear relationships between data points. Support Vector Machines (SVMs) use the kernel trick to find optimal hyperplanes for classification.

Matrix Inversion: Inverting a matrix is used in various contexts, such as solving systems of linear equations and estimating model parameters. Regularization techniques are often applied to prevent ill-conditioning and numerical instability when performing matrix inversion.

Markov Chains: Markov chain transitions can be modeled using transition matrices. These matrices define the probabilities of transitioning between different states over time, making them relevant for various sequence prediction tasks in machine learning.

Graph Representations: Graph data can be represented using adjacency matrices or Laplacian matrices. Matrix operations on these representations facilitate graph-based algorithms like PageRank and community detection.

Optimization: Many optimization algorithms used in machine learning, such as gradient descent and Newton's method, involve matrix operations to update model parameters iteratively and find optimal solutions.

In summary, matrix operations serve as the backbone of numerous machine learning algorithms. They enable algorithms to process and manipulate data efficiently, transforming raw inputs into meaningful representations that can be used for tasks like classification, regression, clustering, and more. The elegance and power of matrix operations lie in their ability to encapsulate complex mathematical transformations, making them an indispensable tool for practitioners working at the forefront of machine learning innovation.

1.3 Vector Spaces and Linear Transformations

Vector spaces and linear transformations form the basis for understanding how data is transformed and manipulated in machine learning. These concepts provide a formal framework for representing data, performing operations, and capturing relationships between variables. Let's explore how vector spaces and linear transformations contribute to the core of machine learning:

Vector Spaces: A vector space is a mathematical structure that encompasses a set of vectors along with operations like addition and scalar multiplication. In machine learning, data often resides in vector spaces, where each vector represents an observation or instance. Vector spaces facilitate the manipulation and transformation of data, allowing algorithms to extract patterns and make predictions.

Feature Transformations: In machine learning, feature transformations involve mapping input features to a different space. Linear transformations play a pivotal role here. They are represented by matrices that define how input vectors are mapped to output vectors. Feature transformations are

used to convert raw data into a format that's more amenable for analysis and modeling.

Basis and Dimension: The concept of a basis in a vector space refers to a set of linearly independent vectors that can be combined to represent any vector in that space. Dimensionality reduction techniques like PCA aim to find a new basis that captures the most important information in the data. By reducing the dimensionality, these techniques aid in visualization, noise reduction, and efficient computation.

Linear Independence and Span: Linear independence is a crucial concept in machine learning. It helps determine the minimal set of features that contribute to prediction, avoiding redundancy. Span refers to the set of all possible linear combinations of a given set of vectors, which defines the subspace they generate. Understanding span is essential for grasping the space in which data resides.

Kernel Trick: The kernel trick is a method to implicitly map data into a higher-dimensional space, enabling linear algorithms to capture complex relationships. This is particularly useful in support vector machines (SVMs) where

linear separation in the transformed space corresponds to nonlinear separation in the original space.

Neural Network Activation Functions: Activation functions in neural networks are a form of nonlinear transformations. They introduce nonlinearity to the model, enabling it to approximate complex functions. These functions introduce intricate relationships between neurons' input and output, allowing neural networks to capture intricate patterns in data.

Image and Audio Transformations: In computer vision and audio processing, linear transformations are applied to images and audio signals. These transformations include resizing, rotating, and scaling images, as well as modifying audio signals for tasks like noise reduction and compression.

Principal Component Analysis (PCA): PCA is a dimensionality reduction technique that involves linear transformations to project data onto a lower-dimensional subspace while preserving its variance. This helps to capture the most significant information in the data, making it a valuable tool in feature extraction and visualization.

Graph Embeddings: In graph-based machine learning, linear transformations are used to embed graphs into vector spaces. These embeddings capture structural information about the nodes and edges, enabling downstream tasks like node classification and link prediction.

Natural Language Processing (NLP): In NLP, linear transformations are employed to map words or phrases to dense vector representations (word embeddings), enabling algorithms to process textual data effectively and capture semantic relationships.

In summary, vector spaces and linear transformations provide a formal framework for understanding how data is transformed, manipulated, and represented in machine learning. They enable algorithms to uncover relationships, reduce dimensionality, and extract valuable features from raw data. The power of these concepts lies in their ability to generalize transformations across data instances, laying the foundation for the rich diversity of techniques and applications within the field of machine learning.

1.4 Eigenvalues and Eigenvectors

Eigenvalues and eigenvectors are essential concepts from linear algebra that have far-reaching implications in machine learning. They offer a unique perspective on data transformations and reveal inherent structures that help algorithms uncover patterns, reduce dimensionality, and make intelligent decisions. Let's explore how eigenvalues and eigenvectors play a crucial role in the machine learning landscape:

Definition and Interpretation: In a nutshell, eigenvectors are non-zero vectors that, when transformed by a matrix, retain their direction. The corresponding scalar values are the eigenvalues. This might seem abstract, but it has profound implications. Eigenvectors capture the most stable directions of transformation, while eigenvalues quantify the scaling factor along those directions.

Dimensionality Reduction: Principal Component Analysis (PCA) is a dimensionality reduction technique that leverages eigenvalues and eigenvectors. By finding the eigenvectors of

the covariance matrix of data, PCA identifies the most significant directions of variation. These eigenvectors become new coordinate axes, and the corresponding eigenvalues represent the amount of variance explained along those axes.

Feature Engineering: Eigenvectors can help transform the original features into a new, more meaningful basis. This process can magnify important patterns while suppressing noise. Such feature engineering is especially useful in scenarios where the original features might not capture the underlying relationships effectively.

Image Compression: In image compression, eigenvectors are employed to decompose images into a set of components that retain most of the information. The fewer components used, the more compression is achieved. This process is similar to PCA and highlights the role of eigenvectors in retaining essential features.

Eigenvalue Decomposition: Eigenvalue decomposition breaks down a matrix into its constituent eigenvalues and eigenvectors. This decomposition has applications in various fields, including graph analysis, data clustering, and solving differential equations.

Spectral Clustering: Eigenvectors are used in spectral clustering, a technique that divides data points into groups based on the spectral properties of a matrix derived from the data. By utilizing the eigenvectors of this matrix, spectral clustering can uncover hidden structures and clusters in data.

Kernel PCA: Kernel Principal Component Analysis (Kernel PCA) uses the notion of eigenvectors in a feature space, even if the transformation to that space is nonlinear. This is achieved through kernel functions, which allow linear operations in a higher-dimensional space without explicitly mapping data points to that space.

Eigenvalues in Deep Learning: In neural networks, especially convolutional neural networks (CNNs), the concept of eigenvalues is linked to the geometry of the loss surface during training. Understanding eigenvalues of the Hessian matrix can provide insights into the network's convergence behavior and the effectiveness of optimization techniques.

Physics Simulations: Eigenvectors and eigenvalues are frequently used in simulations, particularly in quantum mechanics and finite element analysis, where they provide

insights into the modes of vibration, energy levels, and stability of physical systems.

In conclusion, eigenvalues and eigenvectors serve as a powerful lens through which machine learning algorithms can uncover underlying structures, patterns, and relationships within data. Their applications span dimensionality reduction, feature engineering, clustering, compression, and even fundamental understanding of neural network behavior. By enabling algorithms to focus on the most relevant information and discard noise, these concepts play a central role in enhancing the efficiency and effectiveness of machine learning processes.

[49]

Chapter 2: Multivariable Calculus

2.1 Partial Derivatives

2.2 Gradients and Jacobian Matrices

2.3 Chain Rule and Higher-Order Derivatives

2.4 Optimization Techniques

2.1 Partial Derivatives

Partial derivatives are fundamental concepts from calculus that play a critical role in optimization, a cornerstone of machine learning. These derivatives allow us to understand how a function changes with respect to a single variable while holding other variables constant. In the context of machine learning, partial derivatives are key components of optimization algorithms, enabling the fine-tuning of models to find optimal solutions. Here's how partial derivatives contribute to the machine learning landscape:

Gradient Descent: Gradient descent is a fundamental optimization technique used to minimize functions iteratively. Partial derivatives of the loss function with respect to model parameters guide the algorithm toward the direction of steepest decrease in the loss. This enables models to adjust their parameters to fit the training data more accurately.

Stochastic Gradient Descent (SGD): Stochastic gradient descent extends gradient descent by using subsets (mini-batches) of the training data. Partial derivatives are calculated for each mini-batch, leading to faster convergence and lower computational requirements. SGD is widely used in training large-scale neural networks and other machine learning models.

Learning Rate Adaptation: The rate at which optimization algorithms converge depends on the learning rate. Partial derivatives provide insights into the slope of the loss function, helping to adapt the learning rate dynamically during optimization. Techniques like AdaGrad, RMSProp, and Adam adjust the learning rate based on the historical gradient information.

Neural Network Training: In deep learning, partial derivatives are crucial for backpropagation, the process by which gradients are computed with respect to each weight and bias in the network. These gradients are then used to update the model's parameters, enabling the network to learn from data.

Optimization Landscapes: Partial derivatives offer insights into the topology of optimization landscapes. By analyzing the gradients, practitioners can identify regions of fast convergence, plateaus, and saddle points. This understanding guides the choice of optimization algorithms and learning rates.

Regularization: Partial derivatives also play a role in regularization techniques like L1 and L2 regularization. These techniques add penalty terms to the loss function to prevent overfitting. The derivatives of these penalty terms affect how model parameters are adjusted during optimization.

Hyperparameter Tuning: Machine learning models often have hyperparameters that influence their performance. Partial derivatives provide insights into how changes in hyperparameters impact the model's behavior and guide the search for optimal hyperparameter settings.

Bayesian Optimization: Bayesian optimization is a technique used to find the optimal input values of a black-box function. Partial derivatives can help in guiding the search process more efficiently by identifying regions of high uncertainty or rapid change.

Reinforcement Learning: In reinforcement learning, partial derivatives are used in policy gradient methods, where the gradient of the expected reward with respect to the policy parameters is computed to optimize the policy.

Natural Language Processing: In NLP, partial derivatives are used in training models like word embeddings and recurrent neural networks (RNNs), which process sequential data and require optimizing over sequences of inputs and outputs.

In essence, partial derivatives empower machine learning algorithms to navigate the complex landscape of optimization. They guide algorithms toward optimal solutions by indicating how parameters should be adjusted to minimize loss functions. As machine learning continues to advance, the understanding and efficient computation of partial derivatives remain crucial for developing effective and robust learning algorithms.

2.2 Gradients and Jacobian Matrices

Gradients and Jacobian matrices are indispensable tools in machine learning, providing insights into how functions change with respect to multiple variables. These concepts are crucial for optimization, sensitivity analysis, and understanding the behavior of complex models. Let's delve into how gradients and Jacobian matrices contribute to the machine learning landscape:

Gradients: The gradient of a function is a vector that points in the direction of the steepest increase of the function. In machine learning, the gradient of a loss function with respect to model parameters indicates how changing each parameter affects the loss. By following the negative gradient direction, optimization algorithms such as gradient descent aim to find optimal parameter values that minimize the loss.

Backpropagation: In neural networks, the gradient of the loss function with respect to the weights and biases is crucial for backpropagation. This technique calculates gradients through the network layers, enabling the adjustment of weights to minimize the error. The chain rule from calculus is used to efficiently compute these gradients layer by layer.

Sensitivity Analysis: Gradients provide insights into the sensitivity of a model's output to changes in input variables. In regression, for example, the gradient of the output with respect to an input variable helps understand how changes in that variable influence the prediction.

Feature Importance: Gradients can also help quantify the importance of different features in a machine learning model. By analyzing the magnitude of gradients with respect to individual features, practitioners can identify which features have the most significant impact on predictions.

Second-Order Gradients (Hessian): Second-order gradients, often represented by the Hessian matrix, provide additional information about the curvature of the function around a point. The Hessian can be used to refine optimization algorithms, providing information about the local shape of the loss surface.

Jacobian Matrices: While gradients deal with scalar functions, Jacobian matrices extend this concept to vector-valued functions. A Jacobian matrix contains partial derivatives of each function output with respect to all input variables. In machine learning, Jacobian matrices are used to capture how multiple outputs of a model change in response to changes in multiple inputs.

Neural Network Jacobians: In neural networks, Jacobian matrices are used to analyze the impact of input variations on output predictions. This analysis can provide insights into the model's sensitivity to different input perturbations.

Image Transformations: Jacobian matrices are relevant in image processing and computer graphics. They help understand how small changes in image pixel values affect transformations like scaling, rotation, and translation.

Optimization Techniques: In optimization methods like Newton's method, the Jacobian matrix is utilized to guide the search for optimal points by considering both first-order and second-order information.

Parameter Uncertainty: In probabilistic models, the Jacobian matrix can be employed to propagate uncertainty from input variables to output predictions. This is particularly relevant in Bayesian approaches.

Sensitivity to Hyperparameters: Jacobian matrices can help assess how changes in hyperparameters affect the behavior of a machine learning model, aiding in the process of hyperparameter tuning.

In conclusion, gradients and Jacobian matrices provide critical insights into how functions change as their inputs change. They empower machine learning algorithms to optimize models efficiently, perform sensitivity analysis, and understand the impact of input variations. Whether in the context of neural networks, optimization, or uncertainty analysis, gradients and Jacobian matrices are essential tools for developing robust and effective machine learning solutions.

2.3 Chain Rule and Higher-Order Derivatives

The chain rule and higher-order derivatives are indispensable concepts from calculus that play a vital role in understanding the relationships and interactions within complex functions. In machine learning, these concepts are crucial for computing gradients, optimizing models, and capturing intricate dependencies. Let's explore how the chain rule and higher-order derivatives contribute to the machine learning landscape:

Chain Rule: The chain rule is a fundamental principle that allows us to compute the derivative of a composite function. In machine learning, where models are composed of layers, activation functions, and transformations, the chain rule provides a systematic way to compute gradients through each layer, enabling efficient backpropagation.

Backpropagation: Backpropagation in neural networks heavily relies on the chain rule. As data propagates through layers, the gradients of the loss function with respect to each layer's output are computed using the chain rule. This process efficiently calculates how changes in the network's parameters affect the final prediction.

Gradient Calculation: The chain rule plays a central role in calculating gradients for functions involving multiple nested operations. When these functions appear in loss functions or optimization objectives, the chain rule helps compute accurate gradients for efficient optimization.

Higher-Order Derivatives: While first-order derivatives capture rates of change, higher-order derivatives reveal how those rates themselves change. Second-order derivatives, like the Hessian matrix, provide insights into the curvature and shape of a function's surface, helping optimization algorithms navigate more effectively.

Newton's Method: Newton's method, an optimization technique, leverages second-order derivatives to find minima or maxima of functions. It converges more quickly in many cases but requires the computation of the Hessian matrix.

Natural Gradient Descent: Natural gradient descent adjusts gradient descent using second-order derivatives, making it less sensitive to the parameterization of models. This

[61]

approach has applications in probabilistic models and optimization in high-dimensional spaces.

Higher-Order Optimization: Optimization algorithms that use higher-order derivatives, such as quasi-Newton methods, leverage information about both first-order and second-order behavior to guide their search for optimal points.

Hessian-Free Optimization: In machine learning, particularly in training neural networks, Hessian-free optimization methods aim to approximate the Hessian matrix without computing it explicitly. This makes optimization more memory-efficient and suitable for large-scale problems.

Adaptive Learning Rates: Techniques like Adam and RMSProp adjust learning rates based on past gradients and second-order statistics, effectively incorporating information from higher-order derivatives for more adaptive optimization.

Model Interpretation: Higher-order derivatives can provide insights into model behavior and sensitivity. Understanding how model outputs change as input values change can aid in model interpretation and decision-making.

Feature Engineering: Higher-order derivatives can inform the choice of feature transformations by revealing how these transformations impact model performance beyond first-order effects.

In summary, the chain rule and higher-order derivatives are essential tools for understanding the behavior of complex functions, optimizing models, and improving the efficiency of optimization algorithms. By enabling efficient computation of gradients through composite operations and providing insights into curvature and shape, these concepts empower machine learning practitioners to navigate intricate landscapes and develop more effective and robust algorithms.

2.4 Optimization Techniques

Optimization techniques are at the heart of machine learning, driving the process of training models to find optimal parameter values that minimize loss functions. These techniques enable algorithms to learn from data, make accurate predictions, and uncover insights hidden within the data. Let's explore how various optimization techniques contribute to the machine learning landscape:

Gradient Descent: Gradient descent is one of the most fundamental optimization techniques. It iteratively updates model parameters in the direction of the negative gradient of the loss function. Variants like batch gradient descent, stochastic gradient descent (SGD), and mini-batch gradient descent enable efficient updates using different subsets of data.

Momentum and Nesterov Accelerated Gradient: These techniques enhance gradient descent by introducing a momentum term. The momentum accumulates gradients from previous steps, leading to faster convergence and escaping local minima.

Adagrad, RMSProp, and Adam: These adaptive optimization methods adjust the learning rate for each parameter based on the historical gradient information. They ensure that frequently updated parameters have smaller learning rates, preventing them from oscillating excessively.

Learning Rate Scheduling: Instead of using a fixed learning rate, this technique adapts the learning rate during training. It can start with a larger learning rate to converge quickly in the beginning and then gradually reduce it for fine-tuning.

Newton's Method: Newton's method utilizes second-order derivatives (Hessian matrix) to find minima or maxima of functions. While powerful, it can be computationally expensive for high-dimensional problems.

Conjugate Gradient: Conjugate gradient is an iterative method for solving large systems of linear equations, often used for quadratic optimization problems in machine learning.

Quasi-Newton Methods: These methods approximate the Hessian matrix to guide optimization. BFGS and L-BFGS are popular quasi-Newton methods used when the full Hessian is difficult to compute.

Hessian-Free Optimization: Particularly useful for training neural networks, this technique approximates the Hessian matrix without calculating it explicitly, making it memory-efficient and suitable for large-scale problems.

Evolutionary Algorithms: These optimization techniques draw inspiration from biological evolution. They use genetic operators like mutation and crossover to evolve a population of candidate solutions over generations, with the fittest solutions surviving.

Bayesian Optimization: This probabilistic approach models the objective function as a Gaussian process. It intelligently selects new points to evaluate the function, efficiently searching for the optimal point.

Genetic Programming: Genetic programming employs evolutionary algorithms to optimize computer programs, allowing for the discovery of algorithms that perform well on a specific task.

Hyperparameter Optimization: Optimization is not limited to model parameters. Hyperparameter optimization techniques like grid search, random search, and Bayesian optimization help find the best hyperparameters for a given model.

Black-Box Optimization: These techniques aim to optimize functions where the internal structure is unknown. They're used when the function is complex and evaluating it is resource-intensive, common in real-world optimization problems.

In summary, optimization techniques are the driving force behind machine learning model training. They ensure that models learn from data effectively, discover underlying patterns, and generalize well to unseen data. The choice of optimization technique depends on the problem's characteristics, data volume, model architecture, and computational resources. The rich array of techniques empowers machine learning practitioners to fine-tune models, make predictions, and tackle complex challenges across various domains.

Chapter 3: Probability and Statistics

3.1 Basic Probability Concepts

3.2 Random Variables and Probability Distributions

3.3 Expectation, Variance, and Covariance

3.4 Maximum Likelihood Estimation

[69]

3.1 Basic Probability Concepts

Probability theory is a foundational framework in machine learning that helps us model uncertainty, quantify randomness, and make informed decisions in the presence of incomplete or noisy data. Understanding basic probability concepts is essential for tasks such as modeling uncertainty, designing probabilistic models, and making predictions. Here's an exploration of how basic probability concepts contribute to the machine learning landscape:

Probability: Probability measures the likelihood of an event occurring. It's a value between 0 and 1, where 0 indicates impossibility and 1 indicates certainty. In machine learning, probability is used to quantify uncertainty and make predictions based on data.

Random Variables: A random variable is a variable whose possible values are outcomes of a random phenomenon. Discrete random variables can take on a countable set of values (e.g., dice rolls), while continuous random variables have a range of possible values within a given interval (e.g., heights or weights).

Probability Distribution: A probability distribution describes the likelihood of different outcomes in a random experiment. It can be represented graphically using probability density functions for continuous variables or probability mass functions for discrete variables.

Expected Value (Mean): The expected value of a random variable is the average of all possible outcomes, each weighted by its probability. It provides a measure of the central tendency of the variable's distribution.

Variance and Standard Deviation: Variance measures the average squared difference between each value and the expected value. Standard deviation is the square root of the variance. They quantify the spread or dispersion of a distribution.

Joint Probability: Joint probability measures the likelihood of two or more events occurring together. In machine learning, joint probabilities are used in tasks like probabilistic graphical models and Bayesian networks.

Conditional Probability: Conditional probability measures the likelihood of an event occurring given that another event has occurred. It's essential for modeling dependencies and relationships between variables.

Bayes' Theorem: Bayes' theorem is a fundamental concept in probability theory that establishes a relationship between conditional probabilities. It's widely used in machine learning for probabilistic reasoning, Bayesian inference, and updating beliefs as new data is observed.

Independence: Two events are independent if the occurrence of one event doesn't affect the likelihood of the other event occurring. Independence assumptions are often made in machine learning models to simplify computations.

Marginalization: Marginalization involves summing or integrating over one or more variables in a joint probability distribution to obtain the distribution of a subset of variables. It's a crucial step in many probabilistic inference tasks.

Maximum Likelihood Estimation (MLE): MLE is a method for estimating the parameters of a probability distribution that maximize the likelihood of the observed data. It's a common technique for fitting models to data.

Bayesian Inference: Bayesian inference combines prior knowledge and observed data to update beliefs about model parameters. It's a probabilistic approach to learning that accommodates uncertainty in both model parameters and predictions.

In summary, basic probability concepts provide a systematic framework for handling uncertainty and randomness in machine learning. From modeling uncertainty to making informed decisions, understanding probability is crucial for developing accurate and robust machine learning models. These concepts form the bedrock upon which probabilistic models, Bayesian reasoning, and uncertainty quantification are built.

3.2 Random Variables and Probability Distributions

Random variables and probability distributions are fundamental concepts in machine learning that enable us to model and analyze uncertain data. They provide a formal framework for understanding the likelihood of different outcomes and play a crucial role in tasks ranging from data generation to statistical inference. Here's a closer look at how random variables and probability distributions contribute to the machine learning landscape:

Random Variables: A random variable is a variable that can take on different values, each associated with a certain probability. In machine learning, random variables are used to represent uncertain or variable quantities, such as the outcome of a dice roll or the price of a stock at a certain time.

Discrete Random Variables: Discrete random variables have a countable set of possible values. Examples include the outcome of a coin flip (heads or tails) or the number of cars passing through an intersection in a given time interval.

Continuous Random Variables: Continuous random variables can take on any value within a certain range. Examples include the height of a person or the temperature in a specific location.

Probability Mass Function (PMF): For a discrete random variable, the probability mass function assigns probabilities to each possible value. It describes the distribution of the random variable's outcomes.

Probability Density Function (PDF): For a continuous random variable, the probability density function specifies the relative likelihood of the variable taking on different values within a range. The area under the PDF over a certain interval corresponds to the probability of the variable falling within that interval.

Cumulative Distribution Function (CDF): The cumulative distribution function gives the probability that a random variable takes on a value less than or equal to a given value. It provides a complete description of the distribution of a random variable.

Expectation (Mean): The expectation, or expected value, of a random variable is a measure of its central tendency. For discrete random variables, it's the weighted average of all possible values, while for continuous random variables, it's calculated using an integral.

Variance and Standard Deviation: Variance quantifies the spread or variability of a random variable's values around its mean. Standard deviation is the square root of the variance. They provide insights into the variability of outcomes.

Probability Distributions: Probability distributions describe how probabilities are distributed over the possible values of a random variable. Common distributions include the Bernoulli distribution, the Normal distribution, the Exponential distribution, and the Poisson distribution.

Multivariate Distributions: Multivariate distributions describe the joint behavior of multiple random variables. They are used to model relationships and dependencies between variables.

Empirical Distributions: Empirical distributions are approximations of probability distributions based on observed data. They are useful for estimating the underlying distribution when the true distribution is unknown.

Modeling Uncertainty: Random variables and probability distributions are crucial for modeling uncertainty in machine learning. They provide a formal way to represent and reason about uncertain quantities, enabling better decision-making.

In summary, random variables and probability distributions are foundational concepts that enable us to model and analyze uncertainty in machine learning. By providing a mathematical framework for understanding the likelihood of different outcomes, these concepts form the basis for probabilistic reasoning, statistical modeling, and data-driven decision-making.

3.3 Expectation, Variance, and Covariance

Expectation, variance, and covariance are key statistical measures that provide valuable insights into the properties and relationships of random variables. In machine learning, these concepts are crucial for understanding data characteristics, building models, and making informed decisions. Let's explore how expectation, variance, and covariance contribute to the machine learning landscape:

Expectation (Mean): Expectation, often referred to as the mean, is a measure of central tendency that represents the average value of a random variable. It provides insights into the typical or "expected" outcome of a random experiment. In machine learning, the expectation is used to summarize data and model predictions.

Discrete Expectation: For a discrete random variable X with probability mass function P(X), the expectation E[X] is calculated as the sum of each possible value of X weighted by its corresponding probability: $E[X] = \Sigma(x * P(x))$.

Continuous Expectation: For a continuous random variable X with probability density function f(X), the expectation E[X] is computed as the integral of x * f(x) over the entire range of X.

Variance: Variance measures the spread or dispersion of a random variable's values around its mean. It quantifies the variability or uncertainty of the random variable. A low variance indicates that values are close to the mean, while a high variance indicates greater variability.

Variance Formula: The variance of a random variable X is calculated as the average of the squared differences between each value of X and its mean, weighted by their probabilities: $Var(X) = E[(X - E[X])^2]$.

Standard Deviation: The standard deviation is the square root of the variance. It provides a more interpretable measure of spread in the same units as the original data.

Covariance: Covariance measures the degree to which two random variables change together. It indicates whether the variables tend to increase or decrease together or move in opposite directions.

Covariance Formula: The covariance between two random variables X and Y is calculated as the expected product of their deviations from their respective means:

Cov(X, Y) = E[(X - E[X]) * (Y - E[Y])].

Correlation: Correlation is a standardized version of covariance that measures the strength and direction of a linear relationship between two variables. It ranges between -1 (perfect negative correlation) and 1 (perfect positive correlation), with 0 indicating no linear correlation.

Independence and Covariance: If two random variables are independent, their covariance is zero. However, a covariance of zero doesn't necessarily imply independence.

Use in Machine Learning: Expectation, variance, and covariance are used in various machine learning tasks. They provide insights into data characteristics, help in feature selection, and are used in algorithms like Principal Component Analysis (PCA) and linear regression.

Decision Making: These concepts are used to quantify uncertainty, assess risk, and make informed decisions in probabilistic settings.

In summary, expectation, variance, and covariance are fundamental statistical measures that play a pivotal role in understanding and analyzing data in machine learning. They provide valuable insights into the characteristics of random variables and their relationships, helping practitioners make sense of uncertainty, variability, and patterns within the data.

3.4 Maximum Likelihood Estimation

Maximum Likelihood Estimation (MLE) is a powerful statistical method used in machine learning to estimate the parameters of a statistical model. It aims to find the parameter values that make the observed data most probable under the assumed model. MLE plays a central role in various machine learning tasks, from fitting models to data to making predictions. Let's explore how Maximum Likelihood Estimation contributes to the machine learning landscape:

Likelihood Function: The likelihood function measures how probable the observed data is for a given set of model parameters. It quantifies the fit between the model and the data, providing a foundation for parameter estimation.

Log-Likelihood: In practice, it's often more convenient to work with the log-likelihood, which is the natural logarithm of the likelihood function. The log-likelihood simplifies computations and helps prevent numerical underflow.

Parameter Estimation: MLE aims to find the values of model parameters that maximize the likelihood (or log-likelihood) of the observed data. In other words, it seeks the parameter values that make the data most likely under the model.

Interpretation: MLE provides estimates of the model parameters that are consistent with the observed data. These estimates reflect the "best-fitting" parameters according to the assumed model.

Gaussian Distribution Example: Consider estimating the mean and variance of a Gaussian (normal) distribution. MLE seeks the mean and variance values that maximize the likelihood of the observed data points being drawn from that Gaussian distribution.

Inference: MLE estimates serve as the foundation for statistical inference. Confidence intervals and hypothesis tests can be built around MLE estimates to make statements about the true parameter values.

Optimization: Finding the parameter values that maximize the likelihood often involves optimization techniques.

Gradient-based methods, such as gradient descent, are commonly used to maximize the log-likelihood.

Overfitting: In some cases, MLE can lead to overfitting if the model is too complex and the likelihood is maximized at parameter values that capture noise in the data. Regularization techniques can help mitigate this issue.

Assumptions: MLE assumes that the observed data is generated by the assumed model, and the data points are independent and identically distributed (i.i.d.).

Multivariate MLE: MLE can be extended to estimate parameters for multivariate distributions, such as multivariate Gaussians, where the goal is to find parameters that maximize the joint likelihood of observed multivariate data.

Invariance Property: MLE estimates are not invariant to transformations of the parameter space. For example, if the parameterization changes, the MLE estimates will also change.

Bayesian Connection: In Bayesian inference, the MLE estimates can serve as the point estimates for the parameters, forming a connection between frequentist and Bayesian methods.

In summary, Maximum Likelihood Estimation is a fundamental technique in machine learning for estimating model parameters from observed data. It forms the basis for fitting models to data, enabling practitioners to make data-driven decisions and draw insights from the underlying distributions that generate the data. MLE's wide applicability and solid theoretical foundation make it a cornerstone in statistical modeling and machine learning.

[86]

Chapter 4: Information Theory

4.1 Entropy and Information Gain

4.2 Kullback-Leibler Divergence

4.3 Mutual Information and Applications

4.1 Entropy and Information Gain

Entropy and information gain are essential concepts used in decision trees and other machine learning algorithms to quantify uncertainty, make informed splits, and guide feature selection. They play a crucial role in creating effective models that can efficiently partition data and make accurate predictions. Let's delve into how entropy and information gain contribute to the machine learning landscape:

Entropy: Entropy is a measure of the impurity or uncertainty in a dataset. In machine learning, it's commonly used to evaluate the "purity" of a set of class labels. The entropy of a set S with respect to a binary classification problem is calculated as:

$$H(S) = -p_1 \log_2(p_1) - p_2 \log_2(p_2)$$

where p1 is the proportion of examples with one class label and p2 is the proportion with the other class label. Entropy is 0, when all examples belong to the same class (perfectly pure) and is maximum when examples are evenly split between classes (maximum uncertainty).

Information Gain: Information gain is the reduction in entropy achieved after splitting a dataset based on a specific attribute. It quantifies how well a particular attribute separates the data into more homogeneous subsets. Information gain for an attribute A with respect to a dataset S is calculated as:

$$IG(S,A) = H(S) - \sum_{v \in values(A)} [|S_v|/|S|] \, H(S_v)$$

where values(A) are the possible values of attribute A, $|S_v|$ is the number of examples in subset S_v and $H(S_v)$ is the entropy of subset S_v.

Decision Trees: Entropy and information gain are used in decision tree algorithms to determine the best attribute to split the data at each node. A split with high information gain means it reduces uncertainty significantly, making it a favorable split.

Gini Impurity: Gini impurity is another measure of impurity used in decision trees. While similar to entropy, it calculates the probability of misclassifying a randomly chosen element from the set.

Categorical and Continuous Attributes: Both entropy and information gain can be extended to handle both categorical and continuous attributes. For continuous attributes, the decision tree algorithm typically involves choosing a threshold value to split the data.

Feature Selection: Information gain is used in feature selection to rank attributes based on their ability to split the data effectively. Attributes with higher information gain are preferred as they contribute more to class separation.

Limitations: While entropy and information gain are powerful tools, they can favor attributes with more values or attributes with high cardinality. They may not perform optimally in cases with skewed class distributions.

In summary, entropy and information gain are integral to decision trees and other machine learning algorithms that involve splitting data based on attributes. These concepts

[91]

help quantify uncertainty and measure the effectiveness of attribute splits, leading to the creation of accurate and interpretable models. While they have their limitations, they remain essential tools for guiding feature selection and aiding in data-driven decision-making.

4.2 Kullback-Leibler Divergence

Kullback-Leibler (KL) Divergence, also known as relative entropy, is a mathematical concept used to quantify the difference or "divergence" between two probability distributions. It provides insights into how one distribution differs from another, making it a valuable tool in various machine learning tasks, such as model comparison, information theory, and probabilistic modeling. Let's explore how Kullback-Leibler Divergence contributes to the machine learning landscape:

Definition: Given two probability distributions P and Q over the same set of events, the KL Divergence from P to Q is defined as:

$$D_{KL}(P||Q) = \sum_x P(x) \log P(x)/Q(x)$$

for discrete distributions, or as an integral for continuous distributions.

Properties:

- KL Divergence is not symmetric: $D_{KL}(P||Q) \ne D_{KL}(Q||P)$
- KL Divergence is non-negative: $D_{KL}(P||Q) \geq 0$
- KL Divergence is zero if and only if P and Q are the same distribution.

Interpretation:

- KL Divergence measures the "cost" in terms of information content when using Q to approximate P. It quantifies the additional amount of information needed to represent events from P using the model Q.

Applications:

- **Model Comparison:** KL Divergence can be used to compare how well a model Q approximates the true data distribution P. Smaller KL divergence indicates a better approximation.
- **Information Theory:** KL Divergence is a key concept in information theory, providing a measure of how much information is lost when approximating one distribution with another.
- **Probabilistic Modeling:** In machine learning, KL Divergence is used in tasks like variational inference, where it quantifies the difference between the true

posterior distribution and the approximating distribution.

Variational Autoencoders (VAEs):

- In VAEs, KL Divergence is used to regularize the latent space. It encourages the learned latent distribution to be similar to a known prior distribution, typically a simple Gaussian distribution.

Limitations:

- KL Divergence is not a true metric, as it doesn't satisfy the triangle inequality.
- KL Divergence is sensitive to changes in the tail of the distributions, which can lead to instability in some cases.

Alternative Measures:

- Jensen-Shannon Divergence: A symmetric version of KL Divergence that has better properties for clustering and measuring similarity between distributions.
- Earth Mover's Distance (Wasserstein Distance): Measures the minimum "work" needed to transform one distribution into another.

In summary, Kullback-Leibler Divergence is a versatile tool for quantifying the difference between probability distributions.

[95]

Its applications range from model comparison to information theory to probabilistic modeling. While it has limitations, its ability to provide insights into the divergence between distributions makes it a valuable asset in machine learning for understanding and quantifying uncertainty and model performance.

4.3 Mutual Information and Applications

Mutual Information: Mutual Information (MI) is a measure that quantifies the amount of information shared between two random variables. It provides insights into the extent to which the knowledge of one variable reduces uncertainty about the other. Mutual Information is a key concept in information theory and has various applications in machine learning.

Definition: Given two discrete random variables X and Y with probability distributions P(X, Y), P(X), and P(Y), the Mutual Information between X and Y is defined as:

$$I(X;Y) = \sum_{x \in X} \sum_{y \in Y} P(x,y) \log \frac{P(x,y)}{P(x) \cdot P(y)}$$

Properties:

- Mutual Information is non-negative: $I(X;Y) \geq 0$
- Mutual Information is symmetric: $I(X;Y) = I(Y;X)$
- Mutual Information is zero if and only if X and Y are independent.

Applications:

- **Feature Selection:** Mutual Information can help identify which features are most informative for predicting a target variable. Higher MI values indicate stronger relationships.
- **Clustering and Dimensionality Reduction:** MI can be used to cluster data or reduce dimensionality by selecting features with high mutual information with the target.
- **Image Registration:** In computer vision, MI is used for image registration to align two images based on the information shared between their pixel intensities.
- **Information Theory:** MI is a foundational concept in information theory, used to quantify the amount of shared information between variables.
- **Reinforcement Learning:** In certain cases, MI can be used to design reward functions that capture relevant dependencies in reinforcement learning tasks.

Limitations:

- Mutual Information doesn't capture nonlinear relationships between variables.
- Mutual Information may be sensitive to the scale and units of the variables.

Normalized Mutual Information:

- To mitigate the scale sensitivity, Normalized Mutual Information (NMI) is used. It normalizes MI by dividing it by the square root of the product of the individual entropies.

Entropy and Mutual Information Relationship:

- Mutual Information can be expressed in terms of entropies as: $I(X;Y)=H(X)+H(Y)-H(X,Y)$

Conditional Mutual Information:

- Conditional Mutual Information (CMI) generalizes MI to include a third variable Z. It measures the information shared between X and Y given Z.

Information Bottleneck Method:

- The Information Bottleneck method uses mutual information to extract relevant information from data while discarding unnecessary details. It's used in unsupervised learning and representation learning.

[99]

In summary, Mutual Information is a versatile measure that captures the amount of information shared between random variables. Its applications in feature selection, dimensionality reduction, and various machine learning tasks make it a valuable tool for understanding dependencies and designing effective algorithms.

[100]

Chapter 5: Linear Regression

5.1 Simple Linear Regression

5.2 Multiple Linear Regression

5.3 Least Squares Estimation

5.4 Regularization Techniques

5.1 Simple Linear Regression

Simple Linear Regression is a foundational technique in machine learning and statistics that models the relationship between two variables using a straight line. It's a powerful method for understanding how changes in one variable (the independent variable) affect changes in another variable (the dependent variable). Simple Linear Regression forms the basis for more complex regression models and provides insights into data relationships. Here's a closer look at how Simple Linear Regression contributes to the machine learning landscape:

Basic Idea: Simple Linear Regression aims to find the best-fitting line (linear equation) that describes the relationship between the independent variable (X) and the dependent variable (Y). This line is determined by two parameters: the slope (m) and the intercept (b).

Equation of a Line: The equation of a straight line is typically represented as Y = mx + b, where m is the slope (how much Y changes for a unit change in X) and b is the intercept (the value of Y when X is zero).

Fitting the Line: The goal of Simple Linear Regression is to find the values of m and b that minimize the difference between the observed Y values and the Y values predicted by the line.

Cost Function: The cost function quantifies the difference between predicted and observed Y values. The most common cost function is the Mean Squared Error (MSE), which measures the average squared difference between predictions and actual values.

Optimization: Minimizing the cost function involves finding the values of m and b that result in the smallest MSE. This is typically achieved using optimization techniques like gradient descent.

Interpretation: The slope (m) of the line indicates how much the dependent variable (Y) changes for a one-unit change in the independent variable (X). The intercept (b) represents the predicted value of Y when X is zero.

Assumptions: Simple Linear Regression assumes that there's a linear relationship between the variables and that the residuals (the differences between observed and predicted Y values) are normally distributed and have constant variance.

Coefficient of Determination (R-squared): R-squared measures the proportion of the variance in the dependent variable that's explained by the independent variable. It ranges from 0 to 1, with higher values indicating a better fit of the regression line to the data.

Use Cases: Simple Linear Regression is used for tasks like predicting sales based on advertising spending, estimating the impact of study time on exam scores, and analyzing how temperature affects energy consumption.

Extensions: Simple Linear Regression serves as the foundation for more complex regression techniques, such as Multiple Linear Regression, Polynomial Regression, and Nonlinear Regression.

Limitations: Simple Linear Regression is suitable when the relationship between variables is approximately linear. It may not capture complex relationships present in real-world data.

[105]

In summary, Simple Linear Regression is a straightforward yet powerful technique for modeling and analyzing relationships between two variables. By fitting a straight line to data, it provides a means of understanding how changes in one variable influence changes in another. This foundational concept forms the basis for more advanced regression models and helps extract insights from data to make informed decisions.

5.2 Multiple Linear Regression

Multiple Linear Regression is an extension of Simple Linear Regression that enables the modeling of relationships between a dependent variable and multiple independent variables. This technique is invaluable for understanding how multiple factors collectively influence an outcome and is widely used in various fields to analyze and predict complex data patterns. Let's explore how Multiple Linear Regression contributes to the machine learning landscape:

Basic Idea: Multiple Linear Regression expands the linear relationship modeled in Simple Linear Regression to include multiple independent variables. The goal is to find the best-fitting linear equation that relates the dependent variable (Y) to the multiple independent variables ($X_1, X_2, ..., X_n$).

Equation of the Line: The equation for Multiple Linear Regression takes the form $Y = b_0 + b_1X_1 + b_2X_2 + ... + b_nX_n$, where each b represents a coefficient (slope) corresponding to the respective independent variable.

Coefficient Interpretation: The coefficients $b_1, b_2, ..., b_n$ represent the change in the dependent variable Y for a one-unit change in the corresponding independent variable, while keeping other variables constant.

Matrix Form: Multiple Linear Regression can be represented in matrix notation as Y = Xβ + ε, where Y is the vector of observed dependent variable values, X is the matrix of independent variable values, β is the vector of coefficients, and ε is the vector of residuals.

Least Squares Estimation: The coefficients are estimated using the least squares method, which minimizes the sum of squared differences between observed and predicted Y values.

Interpretation of R-squared: In Multiple Linear Regression, R-squared still measures the proportion of variance in the dependent variable that's explained by the independent variables. It indicates the goodness of fit of the model.

Assumptions: Assumptions similar to Simple Linear Regression apply, including linearity, independence of errors, constant variance of residuals, and normality of residuals.

Feature Selection: Multiple Linear Regression can help identify which independent variables are most important for predicting the dependent variable. Feature selection techniques, such as forward selection or backward elimination, are used to choose the most relevant variables.

Multicollinearity: Multicollinearity occurs when independent variables are highly correlated. It can affect coefficient interpretations and lead to instability in parameter estimates.

Interaction Terms: Interaction terms allow the model to capture combined effects of independent variables. For example, an interaction term between age and income can account for the fact that age might have a different impact on income at different income levels.

Model Evaluation: Model evaluation involves assessing the fit of the model, examining the significance of coefficients, and checking assumptions through residual analysis.

Use Cases: Multiple Linear Regression is applied in fields like economics (predicting GDP based on various economic indicators), social sciences (studying factors affecting happiness), and engineering (predicting product performance based on multiple variables).

Extensions: Multiple Linear Regression can be extended to include nonlinear relationships through techniques like Polynomial Regression and Generalized Additive Models (GAMs).

In summary, Multiple Linear Regression provides a versatile framework for modeling relationships between a dependent variable and multiple independent variables. By considering the joint effects of several factors, it enables the analysis of complex data patterns and the development of predictive models that capture the interplay of variables. This technique is a cornerstone of predictive modeling and data analysis across various domains.

5.3 Least Squares Estimation

Least Squares Estimation is a fundamental method used in linear regression to find the coefficients that define the best-fitting line or hyperplane to the data. It aims to minimize the sum of the squared differences between the observed dependent variable values and the values predicted by the model. This method is widely used to fit linear models and forms the core of many regression algorithms. Let's delve into how Least Squares Estimation contributes to the machine learning landscape:

Basic Idea: Least Squares Estimation seeks to find the coefficients that minimize the sum of squared differences between the observed dependent variable values (Y) and the values predicted by the linear model.

Residuals: The residuals are the differences between the observed Y values and the values predicted by the linear model. The goal is to minimize the sum of the squared residuals.

Objective Function: The objective is to find coefficients that minimize the objective function, which is typically the sum of squared residuals: $\Sigma(y_i - \hat{y}_i)^2$, where y_i is the observed Y value and $\hat{y}_i$ is the predicted Y value for the i-th data point.

Matrix Formulation: In matrix form, the objective is to find the coefficient vector β that minimizes the squared Euclidean norm of the residual vector: $||Y - X\beta||^2$.

Closed-Form Solution: In the case of linear regression, the coefficients that minimize the objective function can be found analytically using the closed-form solution: $\beta = (X^TX)^{-1}X^TY$.

Gradient Descent: In more complex regression settings or when dealing with large datasets, gradient-based optimization techniques like gradient descent can be used to iteratively update coefficients and minimize the objective function.

Regularization: Regularization techniques like Lasso (L1 regularization) and Ridge (L2 regularization) can be applied to prevent overfitting and control the complexity of the model.

Advantages: Least Squares Estimation provides a straightforward and interpretable way to estimate the coefficients of a linear model. It has a closed-form solution for simple linear regression and is well-understood.

Limitations: Least Squares Estimation can be sensitive to outliers and multicollinearity. Outliers can disproportionately affect the fit, and multicollinearity can lead to unstable coefficient estimates.

Generalization: Least Squares Estimation is not limited to linear regression. It's used in various regression techniques, including Multiple Linear Regression, Polynomial Regression, and Nonlinear Regression.

Model Evaluation: The quality of the fitted model is evaluated using metrics like Mean Squared Error (MSE), R-squared, and cross-validation.

Interpretation: The coefficients obtained through Least Squares Estimation indicate how the dependent variable changes in response to changes in the independent variables.

Use Cases: Least Squares Estimation is applied in fields like economics, social sciences, engineering, and finance for tasks such as predicting outcomes, understanding relationships, and making decisions.

In summary, Least Squares Estimation is a cornerstone of linear regression modeling. By minimizing the sum of squared differences between observed and predicted values, it allows us to find the best-fitting line or hyperplane that describes the data. This technique forms the basis for various regression algorithms and serves as a crucial tool for building predictive models in machine learning and statistics.

5.4 Regularization Techniques

Regularization techniques are essential tools in machine learning for preventing overfitting, improving model generalization, and finding a balance between fitting the training data well and maintaining simplicity. These techniques introduce additional constraints or penalties to the model's optimization process, influencing the values of the learned parameters. Let's delve into how regularization techniques contribute to the machine learning landscape:

Basic Idea: Regularization techniques add a penalty term to the objective function that the model aims to minimize. This penalty discourages the model from fitting the training data too closely and helps prevent overfitting.

L1 Regularization (Lasso): L1 regularization adds the absolute values of the coefficients as a penalty to the objective function. It encourages sparsity by pushing some coefficients to exactly zero, effectively performing feature selection.

L2 Regularization (Ridge): L2 regularization adds the squared values of the coefficients as a penalty. It tends to spread the impact of features more evenly and can shrink coefficients towards zero without making them exactly zero.

Elastic Net: Elastic Net is a combination of L1 and L2 regularization. It balances the sparsity-inducing property of L1 regularization with the coefficient-shrinking effect of L2 regularization.

Benefits of Regularization:

- **Preventing Overfitting:** Regularization helps control model complexity, preventing it from fitting noise in the training data and resulting in more generalized models.
- **Improved Generalization:** Regularized models tend to generalize better to unseen data, as they focus on capturing meaningful patterns rather than memorizing the training data.
- **Handling Multicollinearity:** Regularization can mitigate the issues caused by multicollinearity (high correlation between features) by reducing the impact of correlated features.
- **Hyperparameter Tuning:** Regularization introduces hyperparameters (e.g., regularization strength) that

control the amount of penalty applied. Tuning these hyperparameters is essential to find the right balance between bias and variance.

- ***Regularization Path:*** Regularization techniques often result in a regularization path, which shows how the coefficients change as the regularization strength varies. It's useful for understanding the impact of regularization on feature importance.

Use Cases:

- In linear regression, regularization techniques like Lasso and Ridge are applied to prevent overfitting and improve model stability.
- In feature selection, L1 regularization (Lasso) can be used to identify and retain the most important features while discarding irrelevant ones.
- In machine learning algorithms like logistic regression and support vector machines, regularization plays a similar role in controlling model complexity.

Model Interpretation: Regularization can make models more interpretable by reducing the emphasis on noisy or irrelevant features.

Trade-off: Regularization involves a trade-off between fitting the training data well (low bias) and avoiding overfitting (low variance). The optimal balance depends on the specific problem and data characteristics.

Limitations: While regularization helps prevent overfitting, it might not be suitable for all scenarios. In cases where fitting the training data as closely as possible is essential, strong regularization could lead to underfitting.

In summary, regularization techniques are crucial for achieving models that generalize well to new data by controlling model complexity. By adding penalties to the optimization process, regularization helps strike a balance between capturing data patterns and avoiding overfitting. Techniques like L1 and L2 regularization have become fundamental tools in the machine learning toolbox, enhancing the stability, interpretability, and performance of models in a variety of applications.

Chapter 6: Classification

6.1 Logistic Regression

6.2 Softmax Regression

6.3 Binary vs. Multiclass Classification

6.4 Evaluation Metrics

6.1 Logistic Regression

Logistic Regression is a widely used statistical method in machine learning for modeling the relationship between one or more independent variables and a binary outcome. Despite its name, Logistic Regression is used for classification tasks rather than regression. It's a fundamental technique that provides insights into the probability of a certain event occurring and is the foundation for more complex classification algorithms. Let's dive into how Logistic Regression contributes to the machine learning landscape:

Basic Idea: Logistic Regression models the probability of a binary outcome (0 or 1) as a function of independent variables. It uses the logistic function (sigmoid function) to map the linear combination of variables to a probability value between 0 and 1.

Sigmoid Function: The logistic function is an S-shaped curve that "squashes" the linear combination of variables into the range [0, 1]. It's defined as $p(x) = 1 / (1 + e^{-z})$, where z is the linear combination of variables.

Log Odds: The log-odds (logit) transformation of the probability p(x) is represented as log(p(x) / (1 - p(x))) = z. This log-odds can take any value between negative infinity and positive infinity.

Parameter Estimation: Similar to linear regression, Logistic Regression involves estimating coefficients (weights) for the independent variables. These coefficients determine the influence of each variable on the log-odds.

Maximum Likelihood Estimation: Coefficients are estimated using Maximum Likelihood Estimation (MLE), which maximizes the likelihood of observing the given binary outcomes under the logistic model.

Decision Boundary: The decision boundary is a threshold value that separates the predicted probabilities into two classes. For binary classification, if the predicted probability is above the threshold, the instance is assigned to class 1; otherwise, it's assigned to class 0.

Training: Logistic Regression models are trained using optimization techniques like gradient descent to find the optimal coefficient values.

Regularization: L1 (Lasso) and L2 (Ridge) regularization can be applied to Logistic Regression to prevent overfitting and improve generalization.

Multiclass Logistic Regression: Logistic Regression can be extended to multiclass classification problems using techniques like One-vs-Rest (OvR) or softmax regression.

Use Cases: Logistic Regression is used in various fields, including medical diagnostics (disease prediction), finance (credit risk assessment), marketing (customer churn prediction), and natural language processing (sentiment analysis).

Interpretability: Logistic Regression coefficients can be interpreted to understand the direction and magnitude of the influence of each variable on the probability of the binary outcome.

Advantages: Logistic Regression is simple, interpretable, and computationally efficient. It serves as a baseline model for many classification tasks and can be a good starting point before exploring more complex algorithms.

Assumptions: Logistic Regression assumes that the relationship between the independent variables and the log-odds of the binary outcome is linear, and the errors are independent and identically distributed.

In summary, Logistic Regression is a foundational classification algorithm that models the probability of binary outcomes. By mapping the linear combination of variables to a probability using the logistic function, it provides a powerful tool for understanding and predicting binary events. Despite its simplicity, Logistic Regression remains a valuable technique in the machine learning toolkit, offering insights and predictive capabilities across diverse applications.

6.2 Softmax Regression

Softmax Regression, also known as Multinomial Logistic Regression or Multiclass Logistic Regression, is an extension of Logistic Regression to handle multiclass classification problems. It generalizes the idea of Logistic Regression to scenarios where there are more than two classes. Softmax Regression models the probabilities of each class and provides a way to assign an instance to one of multiple classes while maintaining the probabilistic interpretation. Let's explore how Softmax Regression contributes to the machine learning landscape:

Basic Idea: Softmax Regression models the probabilities of multiple classes as a function of independent variables. It uses the softmax function to convert the raw scores (logits) for each class into normalized probabilities.

Softmax Function: The softmax function takes an input vector of scores (logits) for each class and transforms them into a probability distribution over all classes. It's defined as $p(y = k|x) = e^{z_k} / \Sigma(e^{z_i})$ for each class k, where z is the vector of logits.

Parameter Estimation: Similar to Logistic Regression, Softmax Regression involves estimating coefficients (weights) for the independent variables. Each class has its own set of coefficients.

Maximum Likelihood Estimation: Coefficients are estimated using Maximum Likelihood Estimation (MLE), which maximizes the likelihood of observing the given multiclass outcomes under the softmax model.

Training: Softmax Regression models are trained using optimization techniques like gradient descent to find the optimal coefficient values.

Decision Rule: In Softmax Regression, the class with the highest predicted probability is selected as the predicted class for an instance.

Multiclass Formulation: Softmax Regression is formulated to model the joint probability distribution of all classes. It ensures that the predicted probabilities sum to 1 across all classes.

Cross-Entropy Loss: The loss function used in Softmax Regression is the cross-entropy loss, which measures the difference between the predicted probabilities and the true class labels.

Regularization: Similar to Logistic Regression, Softmax Regression can be regularized using techniques like L1 (Lasso) and L2 (Ridge) regularization.

Use Cases: Softmax Regression is used in scenarios with multiple classes, such as image classification (recognizing objects in images), sentiment analysis (categorizing text sentiment), and handwritting recognition.

Interpretability: The coefficients in Softmax Regression provide insights into how each independent variable affects the likelihood of each class.

Advantages: Softmax Regression handles multiclass classification in a way that's consistent with the probabilistic interpretation of Logistic Regression. It provides a powerful and interpretable approach for multiclass problems.

Assumptions: Softmax Regression assumes that the relationship between the independent variables and the logits for each class is linear, and the errors are independent and identically distributed.

Extensions: Softmax Regression can be extended to handle more complex data and feature representations through techniques like deep learning and neural networks.

In summary, Softmax Regression is a versatile technique for tackling multiclass classification problems. By modeling the probabilities of each class and using the softmax function, it provides a probabilistic framework for assigning instances to multiple classes. This approach helps maintain interpretability and consistency with the concepts of probability and logistic regression while handling complex classification scenarios.

6.3 Binary vs. Multiclass Classification

Binary Classification and Multiclass Classification are two fundamental types of classification problems in machine learning. While both involve assigning instances to predefined classes, they differ in terms of the number of classes and the techniques used to address them. Let's examine the key differences and characteristics of these two scenarios:

Binary Classification:

1. **Number of Classes:** In binary classification, there are exactly two classes or categories. Examples include spam detection (spam or not spam), medical diagnosis (disease present or absent), and sentiment analysis (positive or negative sentiment).

2. **Output:** The output of a binary classification model is typically a single probability score or a binary label (0 or 1) indicating the predicted class for each instance.

3. **Algorithms:** Algorithms commonly used for binary classification include Logistic Regression, Support Vector Machines (SVMs), Decision Trees, Random Forests, and Neural Networks.

4. **Evaluation Metrics:** Common evaluation metrics for binary classification include Accuracy, Precision, Recall (Sensitivity), F1-Score, ROC Curve, and AUC.

5. **One-vs-Rest (OvR):** Some algorithms, like SVMs and Softmax Regression, can be adapted to handle multiclass problems by using the One-vs-Rest (OvR) approach. This involves training multiple binary classifiers, each for one class versus the rest.

Multiclass Classification:

1. **Number of Classes:** In multiclass classification, there are three or more classes or categories. Examples include image recognition (identifying objects in images), language identification (detecting the language of a text), and music genre classification.

2. **Output:** The output of a multiclass classification model is typically a probability distribution over all classes, indicating the likelihood of each class for a given instance.

3. **Algorithms:** Algorithms used for multiclass classification include Softmax Regression, Decision Trees, Random Forests, k-Nearest Neighbors (k-NN), Naive Bayes, and Neural Networks.

4. **Evaluation Metrics:** Evaluation metrics for multiclass classification include Accuracy, Precision, Recall, F1-Score (which can be extended to the multiclass case), Confusion Matrix, and multiclass versions of ROC Curve and AUC.

5. **Softmax Regression:** Softmax Regression is a specific algorithm designed for multiclass classification. It extends the concept of Logistic Regression to handle multiple classes and predicts the probabilities of each class.

6. **One-vs-One (OvO):** Another approach to handling multiclass problems is One-vs-One (OvO), where a binary classifier is trained for every pair of classes. The class that wins the most binary comparisons is selected as the final prediction.

Decision Boundary:

- In binary classification, the decision boundary is a line or surface that separates the two classes.
- In multiclass classification, the decision boundary can be more complex, involving multiple regions for each class. It depends on the specific algorithm used.

Use Cases:

- Binary classification is suitable for scenarios where the problem involves two distinct outcomes.
- Multiclass classification is used when there are multiple distinct categories or classes that instances need to be assigned to.

In summary, the distinction between binary and multiclass classification lies in the number of classes and the techniques used to address them. Binary classification deals with two classes, while multiclass classification handles three or more. The choice between the two depends on the nature of the problem and the available data.

6.4 Evaluation Metrics

Evaluation metrics are essential tools in machine learning for quantifying the performance of predictive models. These metrics help assess how well a model's predictions align with the actual outcomes and provide insights into the model's strengths and weaknesses. Different metrics are used for different types of tasks, such as classification, regression, and clustering. Let's explore some key evaluation metrics and their significance:

Classification Metrics:

1. **Accuracy:** Accuracy measures the proportion of correctly classified instances out of all instances. It's a simple and intuitive metric but can be misleading when classes are imbalanced.

2. **Precision:** Precision calculates the proportion of true positive predictions (correctly predicted positive instances) out of all instances predicted as positive. It emphasizes the quality of positive predictions.

3. **Recall (Sensitivity):** Recall calculates the proportion of true positive predictions out of all actual positive instances. It emphasizes the ability to identify positive instances.

4. **F1-Score:** The F1-Score is the harmonic mean of precision and recall. It balances the trade-off between precision and recall and is particularly useful when classes are imbalanced.

5. **Specificity:** Specificity calculates the proportion of true negative predictions (correctly predicted negative instances) out of all actual negative instances.

6. **Confusion Matrix:** A confusion matrix is a table that shows the counts of true positive, true negative, false positive, and false negative predictions. It's a visual representation of a model's performance.

7. **Receiver Operating Characteristic (ROC) Curve:** The ROC curve plots the true positive rate (recall) against the false positive rate at various threshold values. It helps visualize the trade-off between sensitivity and specificity.

8. **Area Under the ROC Curve (AUC):** AUC measures the area under the ROC curve. It provides a single value to quantify the overall performance of a classifier.

Regression Metrics:

1. **Mean Absolute Error (MAE):** MAE calculates the average absolute difference between predicted and actual values. It gives an idea of the model's average prediction error.

2. **Mean Squared Error (MSE):** MSE calculates the average squared difference between predicted and actual values. It penalizes larger errors more heavily.

3. **Root Mean Squared Error (RMSE):** RMSE is the square root of MSE. It's more interpretable and provides a measure of the average magnitude of errors.

4. **R-squared (Coefficient of Determination):** R-squared measures the proportion of variance in the dependent variable that's explained by the model. It ranges from 0 to 1, with higher values indicating better fit.

Clustering Metrics:

1. **Silhouette Score:** Silhouette Score measures the compactness and separation of clusters. It quantifies how well instances within the same cluster are similar to each other and how well they are different from instances in other clusters.

2. **Inertia:** Inertia measures the sum of squared distances between instances and their cluster's centroid. It provides a sense of how tightly packed clusters are.

3. **Davies-Bouldin Index:** Davies-Bouldin Index assesses the average similarity between each cluster and its most similar cluster, while considering the distance between their centroids.

Choosing Metrics: The choice of evaluation metrics depends on the problem type, data characteristics, and the specific goals of the analysis. It's important to select metrics that align with the desired outcomes and reflect the practical implications of the model's predictions.

[136]

Imbalanced Data: For imbalanced datasets, metrics like precision, recall, and F1-Score are more informative than accuracy, as they account for the correct prediction of minority classes.

In summary, evaluation metrics provide a quantitative way to assess the performance of machine learning models. By analyzing these metrics, practitioners can gain insights into a model's strengths and weaknesses, make informed decisions about model selection and improvement, and ensure that the model aligns with the intended goals of the analysis.

[137]

Chapter 7: Support Vector Machines

7.1 Linear SVM

7.2 Non-linear SVM

7.3 Kernel Trick

7.4 Margin and Slack Variables

7.1 Linear SVM

The Linear Support Vector Machine (SVM) is a foundational algorithm in the realm of machine learning, particularly in the field of classification. It excels in binary classification tasks by finding the optimal hyperplane that best separates two classes of data points while maximizing the margin between them. This technique is built upon the principles of linear algebra, optimization, and convex geometry.

Margin Maximization and Hyperplane: At the heart of Linear SVM is the notion of margin. The margin is the distance between the hyperplane and the nearest data points from either class. SVM aims to find the hyperplane that maximizes this margin, as it is thought to generalize better to unseen data.

Linear Separability: Linear SVM assumes that the data is linearly separable, meaning that there exists a hyperplane that can completely separate the two classes. If data is not linearly separable, SVM introduces a concept of soft margins, allowing for some misclassification while still striving to find the best separating hyperplane.

Support Vectors: The data points that are closest to the hyperplane, those that "support" it, are called support vectors. These points are crucial because they define the margin and have the most influence on the hyperplane's position.

Optimization Objective: Linear SVM is formulated as an optimization problem. It aims to minimize the weights of the hyperplane (to avoid overfitting) while simultaneously maximizing the margin. This optimization problem is typically solved using techniques such as quadratic programming.

Kernel Trick: While Linear SVM works well for linearly separable data, it can be extended to handle nonlinear data through the kernel trick. By transforming the data into a higher-dimensional feature space using a kernel function, SVM can find a hyperplane that effectively separates the data even when linear separation is not possible in the original space.

Regularization: Linear SVM incorporates a regularization parameter (often denoted as C) that balances the trade-off between maximizing the margin and minimizing the classification error. A small C emphasizes the margin, potentially leading to more misclassifications, while a larger C prioritizes accurate classification, potentially leading to a smaller margin.

Applications: Linear SVM finds applications in a wide range of fields such as text classification, image classification, bioinformatics, and finance. It's particularly useful when dealing with high-dimensional data, where other classification methods might struggle.

Advantages: Linear SVM offers robustness against overfitting due to its margin-maximizing nature. It can handle high-dimensional data and is often effective even when the data isn't perfectly linearly separable. Additionally, the decision boundary is defined only by the support vectors, making it less sensitive to outliers.

Limitations: Linear SVM's main limitation is its reliance on linear separability. In cases where data is not linearly separable, or when decision boundaries are highly nonlinear, nonlinear SVMs (using kernel functions) might be more appropriate.

[142]

In summary, the Linear Support Vector Machine is a powerful classification algorithm that leverages the principles of linear algebra and optimization to find the optimal hyperplane for separating data. Its focus on maximizing the margin ensures better generalization to unseen data and contributes to its robustness. While it's particularly suitable for linearly separable data, its extension through the kernel trick allows it to handle more complex scenarios, making it a versatile tool in the machine learning toolkit.

7.2 Non-linear SVM

The Non-linear Support Vector Machine (SVM) is an extension of the linear SVM that enables the classification of data that is not linearly separable in its original feature space. By applying the kernel trick, non-linear SVMs can transform the data into a higher-dimensional space, where it might become linearly separable, thus allowing the SVM to find more complex decision boundaries. This technique plays a crucial role in tackling intricate classification tasks across various domains.

Kernel Trick and Feature Mapping: The kernel trick is the essence of non-linear SVM. It avoids explicitly transforming data into a higher-dimensional space but calculates the dot product between data points in that space. This is achieved through a kernel function, which defines the similarity or distance between data points in the transformed space without explicitly computing the transformation itself.

Kernel Functions: Various kernel functions are used in non-linear SVMs to define the similarity between data points. Common kernel functions include polynomial kernels, radial basis function (RBF) kernels, sigmoid kernels, and more. Each kernel has its strengths and is suitable for different types of data and decision boundaries.

Mapping to Higher Dimensions: The kernel trick allows non-linear SVM to effectively operate in a higher-dimensional feature space without the need to explicitly calculate the coordinates in that space. This is critical in maintaining computational efficiency even when the transformed space is high-dimensional.

Radial Basis Function (RBF) Kernel: The RBF kernel is one of the most popular kernel functions. It maps data points to an infinite-dimensional space, providing a flexible means to capture intricate decision boundaries. RBF kernels are well-suited for scenarios where the relationship between features and the target is highly complex.

Decision Boundaries in Transformed Space: In the transformed space, the decision boundary may appear linear, even if the original feature space's boundary was non-linear. This is because the kernel trick enables the SVM to capture intricate relationships between data points through the dot product.

Hyperparameter Tuning: Non-linear SVMs introduce additional hyperparameters, such as the choice of kernel function and kernel-specific parameters. Careful tuning of these hyperparameters is essential for achieving optimal performance and preventing overfitting.

Applications: Non-linear SVMs find applications in various domains, including image classification, natural language processing, bioinformatics, and more. They are particularly useful when data exhibits complex, non-linear relationships that cannot be captured by linear classifiers.

Advantages: The key advantage of non-linear SVMs is their ability to handle complex decision boundaries that linear classifiers cannot capture. The kernel trick allows them to transform data implicitly into higher-dimensional spaces, making them suitable for a wide range of classification tasks.

Limitations: While non-linear SVMs are powerful, they can be computationally intensive, especially with large datasets and certain kernel functions. Proper selection and tuning of the kernel function and hyperparameters are crucial to avoid overfitting.

[146]

In summary, non-linear SVMs extend the capabilities of the original linear SVM by leveraging the kernel trick to operate effectively in higher-dimensional spaces. By doing so, they address the limitations of linear separability, allowing for the classification of complex and non-linearly separable data. This versatility makes non-linear SVMs an indispensable tool in machine learning, enabling the handling of intricate classification tasks across diverse domains.

7.3 Kernel Trick

The Kernel Trick is a transformative concept in machine learning, particularly in the context of Support Vector Machines (SVMs). It's a mathematical technique that allows algorithms to operate in higher-dimensional feature spaces without explicitly calculating the transformations. This ingenious trick enables SVMs to handle complex, non-linear relationships between data points while maintaining computational efficiency.

Motivation: In many real-world scenarios, data is not linearly separable in its original feature space. However, it might become separable in a higher-dimensional space. The challenge lies in the computational cost of explicitly mapping data into that space, as it involves calculating coordinates in a potentially high-dimensional domain. The Kernel Trick elegantly addresses this issue.

Kernel Function: At the heart of the Kernel Trick is the kernel function. A kernel function calculates the similarity or distance between data points in the transformed space without requiring the explicit calculation of the coordinates in that space. It effectively encapsulates the transformation in a concise manner.

Dot Product in Higher Dimension: The kernel function calculates the dot product of data points in the transformed space. This allows SVMs to work with the kernel values (dot products) instead of the explicit transformations, avoiding the need for heavy computations in the higher-dimensional space.

Common Kernel Functions:

- **Linear Kernel:** The linear kernel computes the standard dot product between data points and is equivalent to the original feature space. It is primarily used in linear SVMs.
- **Polynomial Kernel:** The polynomial kernel raises the dot product to a specified power, introducing polynomial terms to capture non-linear relationships.
- **Radial Basis Function (RBF) Kernel:** The RBF kernel, also known as the Gaussian kernel, assigns weights to data points based on their similarity to a reference point. It's widely used in non-linear SVMs.

- **Sigmoid Kernel:** The sigmoid kernel calculates a function resembling a sigmoid curve. It's commonly used in neural networks and SVMs.

Benefits of the Kernel Trick:

- **Computational Efficiency:** By avoiding explicit transformations, the Kernel Trick prevents the explosion of dimensions while still allowing SVMs to operate in higher-dimensional spaces.
- **Non-linearity:** The Kernel Trick empowers linear algorithms like SVMs to capture complex, non-linear relationships between data points, making them more versatile.

Applications:

- **In SVMs:** The Kernel Trick extends linear SVMs to non-linear classification by effectively transforming data into higher-dimensional spaces where linear separation might be possible.

- **In Other Algorithms:** The Kernel Trick has applications beyond SVMs. It's used in kernelized versions of other algorithms like kernelized Principal Component Analysis (PCA) and kernelized Ridge Regression.

Challenges:

- **Hyperparameter Tuning:** Different kernel functions and their hyperparameters affect performance. Proper selection and tuning are crucial.
- **Computational Cost:** While more efficient than explicit transformations, computing kernel values can still be demanding for large datasets.

In conclusion, the Kernel Trick is a masterful technique that empowers algorithms to harness the power of higher-dimensional spaces without the computational burden of explicit transformations. In SVMs, it's the key to unlocking the capability to handle complex, non-linear data relationships, making it a cornerstone of machine learning's ability to capture intricate patterns in diverse datasets.

7.4 Margin and Slack Variables

In Support Vector Machines (SVMs), the concepts of margin and slack variables play a critical role in finding the optimal balance between precision and flexibility when defining the decision boundary between classes. They allow SVMs to handle cases where data is not perfectly separable and introduce the notion of a soft margin to accommodate misclassified points.

Margin: The margin in an SVM is the distance between the decision boundary (hyperplane) and the nearest data points from both classes. Maximizing the margin is a central principle in SVMs, as it leads to better generalization to unseen data. A wider margin indicates a higher confidence in the classification.

Hard Margin SVM: In a hard margin SVM, the goal is to find a hyperplane that perfectly separates the classes with the largest possible margin. However, hard margin SVMs are sensitive to outliers and noise in the data. If a single data point is an outlier or an error, it can significantly impact the decision boundary.

Soft Margin SVM: To address the limitations of hard margin SVMs, soft margin SVMs introduce the concept of slack variables. Slack variables allow for a certain degree of misclassification in the training data. The objective becomes a trade-off between maximizing the margin and minimizing the misclassification error.

Slack Variables: Slack variables (often denoted as ξ, xi, etc.) are introduced to quantify the degree of misclassification or "slackness" allowed for each data point. They represent the distance between a misclassified point and the decision boundary. The optimization problem in soft margin SVMs involves minimizing the sum of the slack variables while still maximizing the margin.

C Parameter: The parameter C controls the trade-off between maximizing the margin and allowing misclassifications. A smaller value of C emphasizes a wider margin even if some points are misclassified. A larger C allows more points to be misclassified to achieve a narrower margin.

Regularization and Control: The C parameter acts as a form of regularization in SVMs. A high C places more emphasis on precise classification, potentially leading to overfitting if not chosen carefully. A low C emphasizes a broader margin, making the model more resistant to outliers and noise.

Impact of Slack Variables: Slack variables provide SVMs with a level of flexibility to accommodate data that cannot be perfectly separated. They help SVMs handle situations where the data is not completely linearly separable while still aiming for a meaningful margin.

Trade-off and Flexibility: The balance introduced by slack variables addresses the dilemma between model complexity (flexibility) and generalization (precision). By allowing a controlled number of misclassifications, the model can generalize better to unseen data.

Applications: Soft margin SVMs find applications in real-world scenarios where data may not adhere perfectly to linear separability. They are particularly useful when dealing with noisy or overlapping data.

In summary, the concepts of margin and slack variables are integral to Support Vector Machines, especially in cases where data is not linearly separable. The interplay between these concepts allows SVMs to strike a balance between precision and flexibility, enabling them to handle a wide range of classification tasks with varying levels of complexity and noise.

[155]

Chapter 8: Neural Networks and Deep Learning Basics

8.1 Perceptrons and Activation Functions

8.2 Feedforward Neural Networks

8.3 Backpropagation Algorithm

8.4 Training Neural Networks

8.1 Perceptrons and Activation Functions

Perceptrons and activation functions are fundamental components of neural networks, forming the building blocks of complex architectures that underpin the field of deep learning. They play a vital role in transforming input data, introducing non-linearity, and enabling neural networks to model intricate relationships in data.

Perceptrons: Neurons of Neural Networks

- A perceptron is a simplified model of a biological neuron. It takes multiple input signals, applies weights to them, and produces an output signal.
- The weighted sum of inputs, along with a bias term, is passed through an activation function to produce the output.
- Mathematically, the output of a perceptron can be expressed as: Output = Activation(Weighted Sum + Bias)

Activation Functions: Introducing Non-linearity

- Activation functions introduce non-linearity to neural networks, enabling them to approximate complex functions.
- Without non-linearity, neural networks would behave like linear models, severely limiting their capacity to capture intricate patterns.
- Common activation functions include:
- Sigmoid: Maps input to a range between 0 and 1, useful for binary classification.
- ReLU (Rectified Linear Unit): Outputs input for positive values and zero for negative values, avoiding vanishing gradients.
- Tanh (Hyperbolic Tangent): Maps input to a range between -1 and 1, similar to sigmoid but centered at zero.
- Softmax: Used in the output layer for multi-class classification, converts raw scores into probability distributions.

Feedforward Neural Networks: Stacking Perceptrons

- A feedforward neural network is constructed by stacking layers of perceptrons. Each layer transforms input data

using weights and activation functions.
- The input layer receives raw data, hidden layers progressively extract features, and the output layer produces the final prediction.
- The choice of activation functions in each layer depends on the problem's characteristics and network architecture.

Backpropagation: Training Neural Networks

- Backpropagation is the core training algorithm for neural networks. It adjusts weights and biases to minimize the difference between predicted and actual outputs.
- It involves two main steps: forward pass and backward pass. During the forward pass, input data is processed layer by layer. During the backward pass, gradients are computed and used to adjust weights.

Deep Learning Revolution: Stacking Layers for Depth

- The "deep" in deep learning comes from the idea of stacking numerous hidden layers to create deep neural networks.
- Deeper networks can capture hierarchical features and abstract representations, leading to enhanced performance on complex tasks.

Universal Approximation Theorem: Expressive Power

- Neural networks with just a single hidden layer and a non-linear activation function have been proven to be able to approximate any continuous function to arbitrary accuracy.
- This theorem underlines the remarkable expressive power of neural networks.

Applications: Perceptrons and activation functions are pivotal in various domains, including image recognition, natural language processing, speech synthesis, and more.

In summary, perceptrons and activation functions are fundamental to the architecture and capabilities of neural networks. They introduce non-linearity, enabling networks to model intricate relationships in data. The combination of these components, along with training algorithms like backpropagation, has driven the success of deep learning and its ability to tackle a wide range of complex tasks in today's AI landscape.

8.2 Feedforward Neural Networks

Feedforward Neural Networks (FNNs) are the cornerstone of deep learning. These networks consist of interconnected layers of neurons, each layer transforming input data to produce progressively abstract features, ultimately leading to a final output. FNNs are capable of learning intricate patterns and relationships within data, making them highly effective in various tasks such as classification, regression, and more.

Structure of Feedforward Neural Networks:

- FNNs are composed of an input layer, one or more hidden layers, and an output layer.
- Neurons in the input layer correspond to the features of the input data.
- Neurons in hidden layers transform the data using weights and activation functions, capturing increasingly complex representations.
- The output layer produces the final prediction based on the information processed in the hidden layers.

Feedforward Process:

- Input data is fed into the input layer.
- The input is multiplied by weights and bias terms in each neuron, and the result is passed through an activation function.
- The transformed output becomes the input for the next layer.
- This process continues through the hidden layers until the final output is obtained.

Activation Functions:

- Activation functions introduce non-linearity to FNNs, enabling them to approximate complex functions and capture intricate patterns in data.
- Common activation functions include ReLU (Rectified Linear Unit), sigmoid, tanh (hyperbolic tangent), and softmax.

Weight Initialization:

- Proper initialization of weights is essential for efficient training of FNNs.

- Careful initialization helps prevent vanishing or exploding gradients, which can hinder convergence.

Training FNNs:

- FNNs are trained using algorithms like backpropagation and gradient descent.
- Backpropagation computes the gradients of the loss function with respect to the weights, enabling adjustments to minimize the error.
- Gradient descent updates weights in the direction that decreases the loss function.

Overfitting and Regularization:

- FNNs are susceptible to overfitting, where the model fits the training data too closely and performs poorly on unseen data.
- Regularization techniques like dropout and L2 regularization can mitigate overfitting by introducing penalties for large weights or randomly deactivating neurons during training.

Hyperparameters:

- FNNs have various hyperparameters, including the number of hidden layers, the number of neurons per layer, learning rate, batch size, and more.
- Proper tuning of hyperparameters is crucial for achieving optimal model performance.

Deep Learning and Depth:

- The "deep" in deep learning comes from the ability to stack many hidden layers, allowing FNNs to capture intricate hierarchical relationships.
- Deep neural networks have shown remarkable success in image recognition, natural language processing, and other domains.

Applications:

- FNNs are used in diverse applications such as image classification, object detection, text generation, sentiment analysis, and more.

- Their ability to learn complex mappings from data has fueled the advancements in artificial intelligence and machine learning.

In summary, Feedforward Neural Networks are a fundamental architecture in deep learning. They process data layer by layer, progressively capturing intricate patterns and relationships. Activation functions introduce non-linearity, enabling FNNs to model complex functions effectively. The success of deep learning owes much to the depth and versatility of FNNs, which have revolutionized the field and powered remarkable advancements in AI technology.

8.3 Backpropagation Algorithm

The Backpropagation Algorithm is a cornerstone of training neural networks, enabling them to learn from data and adjust their weights to minimize prediction errors. It involves propagating the error gradients backward through the network, allowing each layer to update its weights in a way that reduces the overall loss. Backpropagation plays a crucial role in optimizing neural network performance and enabling them to capture complex patterns in data.

Understanding Backpropagation:

- Backpropagation involves two main steps: the forward pass and the backward pass.
- During the forward pass, input data is processed layer by layer, producing predictions.
- During the backward pass, gradients of the loss function with respect to the weights are computed and used to adjust the weights.

The Chain Rule of Calculus:

- The key mathematical tool behind backpropagation is the chain rule of calculus.
- It allows us to compute how small changes in weights at each neuron affect the final prediction, by propagating gradients backwards through the layers.

Computing Gradients:

- In the output layer, gradients are calculated based on the difference between predicted and actual values.
- Gradients are then propagated backward through each layer, computing how much each neuron's output contributed to the error.

Weight Update:

- Gradients guide the adjustment of weights in the opposite direction of the gradient's slope.
- Learning rate is a hyperparameter that determines the step size of weight updates.

Stochastic Gradient Descent (SGD):

- Backpropagation is often combined with optimization algorithms like stochastic gradient descent.
- In SGD, weights are updated incrementally based on small batches of data, improving convergence speed and resource utilization.

Mini-batch and Batch Gradient Descent:

- Mini-batch gradient descent uses a subset (mini-batch) of the training data in each iteration.
- Batch gradient descent uses the entire training dataset in each iteration.
- Mini-batch strikes a balance between computational efficiency and convergence speed.

Regularization and Weight Updates:

- Regularization techniques like L2 regularization can be incorporated into the gradient computation and weight update process to prevent overfitting.

Variants and Enhancements:

- Various enhancements to the basic backpropagation algorithm have been developed, such as momentum, adaptive learning rates (e.g., Adam optimizer), and learning rate schedules.

Challenges and Considerations:

- Vanishing and exploding gradients can impact the convergence of backpropagation. Activation functions like ReLU help mitigate these issues.
- Proper weight initialization, regularization, and hyperparameter tuning are crucial for effective training.

Applications:

- Backpropagation is at the core of training neural networks for a wide range of applications, including image and speech recognition, natural language processing, autonomous driving, and more.

[170]

In summary, the Backpropagation Algorithm revolutionized the training of neural networks. By efficiently calculating gradients and adjusting weights layer by layer, it allows neural networks to learn complex patterns from data. The algorithm's ability to optimize network parameters has played a pivotal role in the success of deep learning and its impact across various domains.

8.4 Training Neural Networks

Training neural networks involves iteratively adjusting the model's parameters, such as weights and biases, to minimize the difference between predicted and actual outcomes. This process, driven by optimization algorithms, allows neural networks to learn and generalize from data, enabling them to make accurate predictions on new, unseen examples.

Data Preparation:

- Before training, data needs to be preprocessed, including normalization, scaling, and splitting into training, validation, and test sets.
- Data augmentation techniques can also be employed to increase the diversity of training data, especially for image-related tasks.

Initialization:

- Proper initialization of weights and biases is crucial to prevent training issues like vanishing or exploding gradients.

- Common methods include using random initialization or techniques like Xavier/Glorot initialization.

Loss Function:

- The loss function (also known as the cost function) quantifies the difference between predicted and actual outcomes.
- Common loss functions include mean squared error for regression and categorical cross-entropy for classification.

Optimization Algorithms:

- Optimization algorithms update weights and biases to minimize the loss function.
- Stochastic Gradient Descent (SGD) is a common optimization algorithm, with variants like Mini-batch SGD and Batch Gradient Descent.
- Advanced optimizers like Adam, RMSProp, and AdaGrad adapt learning rates and momentum during training.

Learning Rate and Hyperparameter Tuning:

- The learning rate determines the step size of weight updates during optimization.
- Hyperparameter tuning involves selecting optimal values for learning rate, regularization strength, batch size, and more.
- Learning rate schedules can also be employed to dynamically adjust learning rates during training.

Backpropagation and Gradients:

- Backpropagation computes gradients of the loss function with respect to network parameters, indicating how much each parameter should change to minimize the loss.
- Gradients are used to update weights and biases in the opposite direction of the gradient's slope.

Overfitting and Regularization:

- Overfitting occurs when a model performs well on training data but poorly on new, unseen data.

- Regularization techniques like L1 and L2 regularization, dropout, and early stopping help prevent overfitting by introducing penalties for complex models or dropping random neurons.

Validation and Early Stopping:

- Validation data is used to monitor model performance during training without influencing weight updates.
- Early stopping involves halting training when validation performance starts to degrade, preventing overfitting.

Model Evaluation:

- After training, the model's performance is evaluated on a separate test dataset to assess its generalization capability.
- Various metrics like accuracy, precision, recall, F1-score, and Mean Squared Error (MSE) are used depending on the task.

Transfer Learning and Fine-Tuning:

- Transfer learning involves using pre-trained models as a starting point for specific tasks, reducing training time and data requirements.
- Fine-tuning includes adapting pre-trained models to the new task by updating certain layers' weights.

Applications:

- Training neural networks has revolutionized various fields, including image and speech recognition, natural language processing, healthcare, autonomous vehicles, and more.

In summary, training neural networks is a complex yet essential process in deep learning. By iteratively optimizing model parameters, neural networks learn to capture patterns in data and generalize to new examples. Proper data preprocessing, initialization, optimization algorithms, hyperparameter tuning, and regularization are crucial for achieving successful model training and obtaining state-of-the-art performance on a wide range of tasks.

Chapter 9: Convolutional Neural Networks (CNNs)

9.1 Image Representation

9.2 Convolution and Pooling Layers

9.3 CNN Architectures

9.4 Transfer Learning with CNNs

9.1 Image Representation

Convolutional Neural Networks (CNNs) have revolutionized image analysis by introducing a novel approach to representing and understanding images. These networks are designed to automatically learn hierarchical features from raw pixel data, enabling them to capture intricate patterns and structures within images.

Pixel Data to Hierarchical Features:

- Images are represented as matrices of pixel values, with each pixel representing color intensity.
- CNNs transform raw pixel data into a hierarchical representation of features through convolutional, pooling, and fully connected layers.

Local Receptive Fields and Convolution:

- Convolutional layers are the core of CNNs, using filters (also known as kernels) to perform convolutions.
- Each filter scans a small local receptive field of the input image, capturing spatial patterns.

- Convolutions create feature maps that highlight different aspects of the image.

Shared Weights and Feature Detection:

- CNNs employ shared weights in convolutional layers, allowing filters to detect similar patterns across the image.
- Early layers often capture low-level features like edges, corners, and textures.

Pooling for Subsampling:

- Pooling layers reduce spatial dimensions and computational complexity by subsampling feature maps.
- Max pooling, for example, selects the maximum value within a pooling window, retaining important features.

Hierarchical Features and Stacking:

- As images progress through convolutional and pooling layers, the network captures increasingly complex features.

- High-level layers represent abstract concepts like object parts and entire objects.

Global Information through Fully Connected Layers:

- Fully connected layers aggregate information from the hierarchical features to make predictions.
- These layers are used for classification, regression, and other tasks.

Pretrained CNNs and Transfer Learning:

- Pretrained CNNs, trained on large datasets like ImageNet, can extract rich features from images.
- Transfer learning involves fine-tuning a pretrained model for a specific task, using its learned features as a starting point.

Applications:

- CNNs excel in image recognition, object detection, image segmentation, and more.

- They have been applied in self-driving cars, medical image analysis, content recommendation, and art generation.

Advantages:

- CNNs automatically learn features from data, reducing the need for manual feature engineering.
- Hierarchical representation allows them to capture both low-level and high-level image characteristics.

Challenges and Considerations:

- CNNs require substantial computational resources and training data.
- Proper hyperparameter tuning, regularization, and transfer learning techniques are crucial.

In summary, image representation in Convolutional Neural Networks has revolutionized computer vision by enabling networks to automatically learn hierarchical features from raw pixel data. The hierarchical approach allows CNNs to capture intricate patterns and structures in images, making them highly effective in a wide range of image-related tasks.

9.2 Convolution and Pooling Layers

Convolutional and pooling layers are fundamental components of Convolutional Neural Networks (CNNs), enabling these networks to capture and process spatial information in images efficiently. These layers play a crucial role in transforming raw pixel data into hierarchical feature representations, enabling CNNs to learn and extract complex patterns from images.

Convolutional Layers:

- Convolutional layers apply filters (kernels) to local receptive fields of the input image.
- Filters detect specific features such as edges, textures, and corners.
- Convolution involves element-wise multiplication and summation of the filter and receptive field values.

Stride and Padding:

- The stride determines the step size of the filter's movement across the input.

- Padding involves adding extra pixels around the input, which maintains spatial dimensions after convolution.

Shared Weights:

- One of the key features of convolutional layers is weight sharing.
- Filters are reused across the entire input image, allowing the network to detect the same features in different spatial locations.

Feature Maps:

- Each filter produces a feature map that highlights the presence of a specific feature in the input.
- Multiple filters in a convolutional layer create a set of feature maps, each capturing a different aspect of the input.

Pooling Layers:

- Pooling layers downsample feature maps, reducing spatial dimensions and computational load.

- Max pooling selects the maximum value within a pooling window, preserving the most salient features.
- Average pooling takes the average value within the window.

Subsampling and Invariance:

- Pooling introduces spatial invariance, allowing the network to detect features regardless of their precise location in the input.
- It enhances the network's resistance to small translations and distortions.

Stride in Pooling Layers:

- Pooling layers also have a stride parameter, controlling the distance the pooling window moves.
- A larger stride reduces the overlap of pooling windows and further reduces spatial dimensions.

Pooling for Feature Reduction:

- Pooling reduces the dimensionality of the feature maps, decreasing the number of parameters and improving computational efficiency.

- However, excessive pooling can lead to loss of fine-grained spatial information.

Skip Connections and Variants:

- Skip connections, introduced in architectures like ResNet, alleviate the vanishing gradient problem and enable the training of very deep networks.
- Variants of pooling, such as global average pooling, aggregate information globally instead of using local windows.

Applications:

- Convolutional and pooling layers are foundational in image recognition, object detection, image segmentation, and more.
- They are also applied in non-image tasks like time series analysis and natural language processing.

Advantages:

- Convolutional layers capture local patterns efficiently, and pooling layers reduce dimensionality and enhance invariance.

Considerations:

- Stride, padding, and filter size impact the spatial dimensions and computational load.
- Careful tuning of hyperparameters is essential for optimal model performance.

In summary, convolutional and pooling layers are essential building blocks in Convolutional Neural Networks, enabling the networks to process and extract hierarchical features from images. Convolutional layers detect local patterns, while pooling layers downsample and reduce feature dimensions. These layers are instrumental in the success of CNNs across various computer vision tasks.

9.3 CNN Architectures

Over the years, several groundbreaking Convolutional Neural Network (CNN) architectures have been developed, each with its unique design and capabilities. These architectures have achieved remarkable success in various computer vision tasks, setting new benchmarks and pushing the boundaries of what's possible in image analysis.

LeNet-5:

- LeNet-5, designed by Yann LeCun in the 1990s, was one of the earliest successful CNN architectures.
- It was developed for handwritten digit recognition and consisted of convolutional and pooling layers followed by fully connected layers.
- LeNet-5 laid the foundation for modern CNNs and demonstrated the power of learned hierarchical features.

AlexNet:

- AlexNet, introduced by Alex Krizhevsky in the 2012 ImageNet Large Scale Visual Recognition Challenge (ILSVRC), marked a breakthrough.

- It consisted of multiple convolutional and pooling layers with ReLU activations, followed by fully connected layers.
- AlexNet's depth and architecture led to a significant reduction in error rates, setting the stage for deep CNNs.

VGGNet:

- The Visual Geometry Group (VGG) introduced the VGG architecture, known for its simplicity and uniformity.
- VGGNet had a deep structure with a large number of convolutional layers (up to 19 layers) and small 3x3 filters.
- It demonstrated the importance of depth and paved the way for deeper networks.

GoogLeNet (Inception):

- GoogLeNet, developed by researchers at Google, introduced the concept of "Inception" modules, which used multiple filter sizes within a single layer.
- This architecture significantly reduced the number of parameters while maintaining network depth and performance.
- GoogLeNet won the 2014 ILSVRC and showcased the advantages of parallel processing.

ResNet (Residual Networks):

- ResNet, also by researchers at Microsoft, introduced residual connections to address vanishing gradient issues.
- Residual connections allow information to bypass certain layers, enabling training of extremely deep networks (e.g., ResNet-152).
- ResNet's skip connections revolutionized the training of deep networks and became a standard architecture.

DenseNet (Densely Connected Convolutional Networks):

- DenseNet, proposed in 2017, introduced dense connections where each layer is connected to every other layer in a feedforward fashion.
- These connections encouraged feature reuse and enhanced gradient flow, improving information flow across layers.
- DenseNet achieved state-of-the-art performance with fewer parameters.

EfficientNet:

- EfficientNet, introduced in 2019, focuses on achieving high performance with constrained resources.

- It uses a compound scaling method to balance network depth, width, and resolution to optimize efficiency and accuracy.
- EfficientNet models have become popular for mobile and resource-constrained applications.

Applications and Impact:

- These architectures have been applied in various tasks like image classification, object detection, image segmentation, and more.
- They have led to remarkable advances in areas like medical imaging, self-driving cars, and remote sensing.

Advancements and Beyond:

- The field of CNN architectures continues to evolve, with architectures designed for specific tasks (e.g., U-Net for image segmentation) and research on efficient model designs.

In summary, CNN architectures have undergone significant evolution, from early pioneers like LeNet-5 to modern giants

[191]

like EfficientNet. These architectures have demonstrated the power of deep learning in computer vision, leading to groundbreaking advancements and reshaping the landscape of artificial intelligence.

9.4 Transfer Learning with CNNs

Transfer learning is a powerful technique that leverages the knowledge learned from one task and applies it to a different but related task. In the context of Convolutional Neural Networks (CNNs), transfer learning involves using pretrained models on large datasets as a starting point for training on a new, often smaller dataset. This approach allows for faster convergence, improved generalization, and better results, even with limited data.

Motivation for Transfer Learning:

- Neural networks, especially CNNs, require a substantial amount of data for effective training.
- In many real-world scenarios, collecting a large labeled dataset is challenging or expensive.
- Transfer learning mitigates the need for massive datasets by reusing knowledge from existing models.

Steps in Transfer Learning:

- Pretraining: A CNN is pretrained on a large dataset, often using a general task like image classification.
- Feature Extraction: The pretrained CNN's feature extraction layers are retained while discarding the fully connected layers.
- Fine-Tuning: The retained layers are fine-tuned using a smaller dataset relevant to the specific task.

Advantages of Transfer Learning:

- Feature Learning: Pretrained models have already learned a rich hierarchy of features from a diverse dataset.
- Faster Convergence: Starting with pretrained weights allows the model to converge faster on the new dataset.
- Better Generalization: Transfer learning enables the model to generalize better to new data, even with limited samples.
- Resource Efficiency: It saves computational resources and training time.

Scenarios for Transfer Learning:

- Small Datasets: When the target dataset is small, transfer learning helps prevent overfitting.
- Similar Tasks: If the source and target tasks share similarities, knowledge transfer is more effective.
- Different Domains: Even in different domains, lower layers often learn low-level features that can be useful.

Fine-Tuning Considerations:

- Layer Selection: Layers closer to the input usually capture low-level features, while deeper layers capture more abstract features.
- Learning Rate: Lower learning rates are often used for fine-tuning to prevent drastic changes to pretrained features.
- Freezing Layers: Some layers may be frozen to preserve learned features during fine-tuning.

Popular Pretrained Models:

- ImageNet Pretrained Models: Models pretrained on the ImageNet dataset (e.g., VGG, ResNet, Inception) are commonly used for transfer learning.

- BERT and GPT: In the field of natural language processing, pretrained models like BERT and GPT have become influential.

Applications:

- Transfer learning is widely used in various domains, including image classification, object detection, sentiment analysis, and medical imaging.

Limitations:

- Domain Differences: If the source and target domains are vastly different, transfer learning might not yield significant benefits.
- Catastrophic Forgetting: Fine-tuning can sometimes lead to forgetting the knowledge learned from the source task.

Ethical Considerations:

- Ethical considerations arise when transferring models trained on one dataset to new datasets, as biases present in the source data might carry over.

In summary, transfer learning with CNNs is a valuable technique that allows pretrained models to be repurposed for new tasks with limited data. By leveraging the knowledge embedded in existing models, transfer learning offers a pathway to achieving higher performance and better generalization in various applications, contributing to the success of deep learning in real-world scenarios.

Chapter 10: Recurrent Neural Networks (RNNs)

10.1 Sequential Data

10.2 RNN Architecture

10.3 Long Short-Term Memory (LSTM) Networks

10.4 Applications in NLP

10.1 Sequential Data

Sequential data refers to data where the order of elements matters, and each element is influenced by the preceding elements in the sequence. Examples of sequential data include time series, natural language sentences, DNA sequences, and more. Recurrent Neural Networks (RNNs) are designed to effectively handle such data by capturing and utilizing the temporal dependencies present in the sequence.

Challenges of Sequential Data:

- Sequential data often exhibits complex patterns and relationships that evolve over time.
- Traditional machine learning models struggle to capture these temporal dependencies and patterns.

Temporal Dependency and RNNs:

- RNNs are uniquely suited for handling sequential data due to their inherent ability to maintain memory of previous inputs in the sequence.

- They process each element in the sequence while maintaining an internal hidden state that encodes past information.

Recurrent Neurons:

- The key to RNNs is the recurrent neuron, which processes the current input along with the previous hidden state to produce an output and update the hidden state.
- This hidden state acts as a form of memory, allowing the network to remember past information.

Vanishing and Exploding Gradients:

- RNNs can suffer from vanishing and exploding gradient problems during training.
- Long sequences can lead to gradients that either become too small (vanishing) or too large (exploding), making learning difficult.

Long Short-Term Memory (LSTM) Networks:

- LSTM networks are a type of RNN designed to alleviate the vanishing gradient problem and capture longer-range dependencies.
- LSTMs use a combination of input, forget, and output gates to control the flow of information in and out of the memory cell.

Gated Recurrent Units (GRUs):

- GRUs are another variant of RNNs that combine some aspects of LSTMs while simplifying the architecture.
- They have fewer parameters and can be computationally more efficient than LSTMs.

Bidirectional RNNs:

- Bidirectional RNNs process sequences in both directions (forward and backward) and combine the information from both directions.

- This approach is beneficial when the current element in the sequence depends on both past and future elements.

Applications:

- RNNs have a wide range of applications, including natural language processing (language modeling, machine translation, sentiment analysis), speech recognition, music generation, and time series prediction.

Limitations:

- RNNs can struggle with capturing very long-range dependencies due to vanishing/exploding gradients.
- Training RNNs can be computationally intensive, especially for deep architectures.

Future Directions:

- Advances like attention mechanisms, Transformers, and self-attention have expanded the capabilities of modeling sequential data beyond traditional RNNs.

In summary, sequential data is a fundamental type of data where the order of elements matters. RNNs are specialized neural networks that excel at capturing temporal dependencies in sequential data, making them invaluable for tasks involving time series, text, and other ordered data. However, challenges like vanishing gradients have led to the development of advanced architectures like LSTMs and GRUs, and the field continues to evolve with new techniques and models.

10.2 RNN Architecture

Recurrent Neural Networks (RNNs) are a class of neural networks designed to handle sequential data by capturing and utilizing temporal dependencies. Unlike traditional feedforward neural networks, RNNs have connections that loop back on themselves, allowing them to maintain memory of past inputs and process sequences effectively.

Basic Structure:

- At each time step, an RNN takes an input (usually a vector representing an element in the sequence) and combines it with the previous hidden state to produce an output and update the current hidden state.
- The hidden state serves as a form of memory that encodes information from previous steps.

Mathematical Formulation:

- Mathematically, an RNN's hidden state update at time step t can be represented as:

[205]

$h_t = f(W_{hx} x_t + W_{hh} x_{t-1} + b_h)$

where:

- x_t is the input at time step t.
- h_t is the hidden state at time t.
- W_{hx} is the weight matrix for the input.
- W_{hh} is the weight matrix for the hidden state.
- b_h is the bias term,
- f is the activation function (commonly tanh or ReLU).

Unrolling Over Time:

- RNNs are often visualized as unrolled sequences, where the network unfolds in time steps.
- Each unrolled time step represents the same set of weights and biases being reused.

Vanishing Gradient Problem:

- A major challenge with RNNs is the vanishing gradient problem, where gradients become too small to effectively update the weights during backpropagation through time.

- This issue limits the ability of RNNs to capture long-range dependencies.

Long Short-Term Memory (LSTM):

- LSTM networks were introduced to address the vanishing gradient problem and better capture long-range dependencies.
- LSTMs use gates (input, forget, and output) to control the flow of information in and out of a memory cell.
- This architecture enables LSTMs to store and retrieve information over long sequences.

Gated Recurrent Units (GRUs):

- GRUs are a simplified version of LSTMs that combine some of their gating mechanisms.
- They have fewer parameters and can be computationally more efficient.

Bidirectional RNNs:

- In bidirectional RNNs, the input sequence is processed in both the forward and backward directions.
- This allows the network to capture information from both past and future elements.

Applications:

- RNNs are widely used in natural language processing for tasks like language modeling, machine translation, and sentiment analysis.
- They are also applied in speech recognition, time series forecasting, and more.

Limitations:

- RNNs can struggle with long-range dependencies and vanishing gradients.
- More advanced architectures like Transformers have gained popularity for some tasks.

Future Directions:

- The field continues to evolve with techniques like attention mechanisms, self-attention, and hybrid models that combine RNNs with other architectures.

In summary, the architecture of Recurrent Neural Networks revolves around their ability to maintain hidden states over time and capture sequential patterns. While basic RNNs have limitations like vanishing gradients, variants like LSTMs and GRUs address these challenges and have enabled the modeling of more complex temporal dependencies.

10.3 Long Short-Term Memory (LSTM) Networks

Long Short-Term Memory (LSTM) networks are a type of recurrent neural network (RNN) designed to address the vanishing gradient problem and capture long-range dependencies in sequential data. LSTMs have proven to be highly effective in a wide range of tasks involving sequences, such as natural language processing, speech recognition, and time series prediction.

Motivation for LSTMs:

- Traditional RNNs suffer from the vanishing gradient problem, where gradients diminish as they propagate back through time, leading to difficulties in learning long-term dependencies.
- LSTMs were introduced to address this issue by incorporating a memory cell and various gating mechanisms.

Components of LSTM:

- Cell State (Ct): The cell state acts as a conveyor belt that carries information across time steps. It can store long-term information and pass it through time.
- Input Gate (i): The input gate controls how much new information is added to the cell state. It's influenced by the current input and the previous hidden state.
- Forget Gate (f): The forget gate determines what information from the previous cell state should be forgotten. It considers the current input and the previous hidden state.
- Output Gate (o): The output gate controls the amount of information that is output from the cell state to the hidden state and to the next time step.

LSTM Operation:

- The input gate decides which values will be updated in the cell state based on the new input and the previous hidden state.
- The forget gate determines what information to discard from the cell state based on the new input and the previous hidden state.

- The cell state is updated by combining the input from the input gate and the values retained through the forget gate.
- The output gate controls the output of the cell state and the hidden state for the current time step.

Advantages of LSTMs:

- Long-Term Dependencies: LSTMs can capture dependencies over long sequences, making them suitable for tasks with intricate temporal relationships.
- Mitigating Vanishing Gradients: LSTMs use gating mechanisms to regulate the flow of information, mitigating the vanishing gradient problem.
- Less Forgetfulness: The forget gate enables LSTMs to selectively retain important information from previous time steps.

Applications:

- Natural Language Processing: LSTMs are used for language modeling, machine translation, text generation, and sentiment analysis.

- Speech Recognition: LSTMs excel in speech recognition tasks by modeling phonetic and prosodic patterns.
- Time Series Prediction: LSTMs are effective in predicting future values in time series data.

Variants and Enhancements:

- Gated Recurrent Units (GRUs): GRUs are simpler variations of LSTMs that combine some of the gating mechanisms while reducing the number of parameters.
- Peephole Connections: These allow the gates to consider the cell state directly, enhancing the modeling capabilities of LSTMs.

Limitations and Considerations:

LSTMs can still struggle with very long-range dependencies.

Training LSTMs can be computationally intensive, especially for deep architectures.

In summary, Long Short-Term Memory (LSTM) networks have revolutionized the processing of sequential data by effectively

[212]

capturing long-term dependencies and mitigating the vanishing gradient problem. Their ability to remember and forget information over extended sequences makes them a cornerstone of modern deep learning architectures for tasks involving sequences, playing a crucial role in various applications across diverse domains.

10.4 Applications in NLP

Natural Language Processing (NLP) involves the interaction between computers and human language. Sequential models, such as Recurrent Neural Networks (RNNs) and their variants, have had a profound impact on various NLP tasks, enabling machines to understand, generate, and process human language.

Language Modeling:

- Language modeling involves predicting the next word in a sequence given the previous words.
- RNNs, LSTMs, and GRUs are used to model the sequential nature of text, capturing word dependencies and probabilities.

Machine Translation:

- RNNs, especially sequence-to-sequence models, are used for machine translation.

- The encoder-decoder architecture encodes the source sentence and decodes it into the target language.

Text Generation:

- RNNs and LSTMs are employed to generate coherent and contextually relevant text.
- They are used for creative writing, chatbots, and even generating code or music.

Sentiment Analysis:

- Sentiment analysis aims to determine the sentiment or emotion expressed in a piece of text.
- RNNs can be used to capture the sentiment context by considering the entire sequence.

Named Entity Recognition (NER):

- NER involves identifying and classifying entities (names, locations, dates, etc.) in text.

- RNNs can be used to model the context and dependencies around entities.

Part-of-Speech Tagging:

- Part-of-speech tagging assigns grammatical categories (noun, verb, adjective, etc.) to each word in a sentence.
- RNNs help capture contextual information for accurate tagging.

Text Summarization:

RNNs can be used for abstractive or extractive text summarization, condensing long texts into shorter summaries.

Question Answering:

- RNNs can power question-answering systems, where the model answers questions based on a given context.

Language Translation and Generation with Transformers:

- Transformers, a revolutionary model architecture, have become pivotal in NLP.
- They excel in tasks like machine translation (e.g., the famous model "Transformer"), text generation (e.g., GPT-3), and more.

Ethical Considerations:

- With powerful language models, concerns arise regarding potential biases, misuse, and the creation of fake content.

Multilingual and Cross-Lingual Models:

- Transfer learning and pretrained models like multilingual BERT are employed to perform NLP tasks across multiple languages.

Future Directions:

- NLP continues to advance with models like GPT-4, fine-tuning strategies, and addressing ethical challenges.

In summary, NLP has been transformed by the application of RNNs and related models, enabling machines to understand, generate, and manipulate human language. These models have paved the way for major advancements in various NLP tasks, bringing us closer to natural and contextually aware interactions with machines.

[218]

Chapter 11: Unsupervised Learning: Clustering and Dimensionality Reduction

11.1 K-Means Clustering

11.2 Hierarchical Clustering

11.3 Principal Component Analysis (PCA)

11.4 t-Distributed Stochastic Neighbor Embedding (t-SNE)

11.1 K-Means Clustering

K-Means Clustering is a widely used unsupervised machine learning technique designed to uncover hidden patterns or groupings within data. It is particularly useful for exploratory data analysis, customer segmentation, image compression, and various other applications where finding inherent structures within data is essential. K-Means seeks to partition a dataset into a predetermined number of clusters, with each cluster represented by its centroid. Here's a closer look at K-Means Clustering:

Algorithm Overview:

- Initialization: K initial centroids are randomly selected from the data points or determined through other methods.
- Assignment: Each data point is assigned to the nearest centroid, forming clusters.
- Update: The centroids of the clusters are updated by computing the mean of all data points assigned to each cluster.

- Repeat: Steps 2 and 3 are repeated until the centroids converge or a maximum number of iterations is reached.

Key Concepts:

- Distance Metric: The choice of distance metric (usually Euclidean distance) determines how data points are assigned to clusters based on their proximity to centroids.
- Number of Clusters (K): The number of clusters is a critical parameter. It's often determined through domain knowledge, experimentation, or methods like the elbow method.

Advantages:

- Interpretability: K-Means produces easily interpretable results, as each cluster is defined by its centroid.
- Scalability: K-Means can handle large datasets and is relatively fast, making it suitable for a wide range of applications.
- Simplicity: The algorithm's simplicity makes it accessible and efficient, especially for initial data exploration.

Limitations:

- Number of Clusters: Choosing the optimal number of clusters can be challenging. Incorrect K values may lead to suboptimal results.
- Sensitivity to Initialization: K-Means can converge to local minima, resulting in different outcomes depending on the initial centroids.
- Assumption of Equal Variance: K-Means assumes that clusters have equal variance, which may not hold in all cases.
- Non-Globular Clusters: It struggles with identifying clusters with irregular shapes or those with varying densities.

Applications:

- Market Segmentation: K-Means is commonly used to segment customers into distinct groups based on their purchasing behavior, allowing businesses to tailor marketing strategies.
- Image Compression: K-Means can be applied to compress images by representing similar colors with a single color, reducing storage space.

- Anomaly Detection: Detecting outliers or anomalies can be achieved by considering points that are far from any cluster's centroid.
- Text Clustering: K-Means is used to cluster similar documents, enabling document categorization and topic analysis.
- Genomic Data Analysis: In bioinformatics, K-Means is employed to cluster genes based on expression patterns for biological insights.

Extensions and Variants:

- K-Means++: A smart initialization technique that selects initial centroids to improve convergence speed and final results.
- Mini-Batch K-Means: A variation that uses mini-batches of data points to update centroids, making it more scalable for large datasets.
- Hierarchical K-Means: Combines K-Means with hierarchical clustering to build a tree-like structure of clusters.
- Fuzzy C-Means: Extends K-Means by allowing data points to belong to multiple clusters with varying degrees of membership.

[224]

In conclusion, K-Means Clustering is a versatile and widely adopted technique for uncovering patterns and structure within data. Its simplicity, interpretability, and scalability make it a valuable tool for initial data exploration, segmentation, and various clustering tasks across different domains. However, understanding its limitations and choosing appropriate parameters are key to obtaining meaningful and accurate results.

11.2 Hierarchical Clustering

Hierarchical Clustering is another powerful unsupervised machine learning technique used for grouping data points into clusters. Unlike K-Means, Hierarchical Clustering doesn't require specifying the number of clusters beforehand. Instead, it creates a hierarchical structure of clusters that can be visualized as a tree-like diagram (dendrogram). This technique is particularly useful when data exhibits nested or hierarchical relationships. Let's explore the intricacies of Hierarchical Clustering:

Algorithm Overview:

- Initialization: Each data point starts as its own cluster.
- Merge or Agglomerate: Similar clusters are merged iteratively based on a chosen linkage criterion (e.g., single linkage, complete linkage, average linkage) that defines how to measure similarity between clusters.
- Dendrogram: As clusters are merged, a dendrogram is constructed, showing the hierarchy of cluster formations.

Linkage Criteria:

- Single Linkage: Measures the distance between the closest points of two clusters. Prone to "chaining" small clusters together.
- Complete Linkage: Measures the distance between the farthest points of two clusters. More robust to outliers and "chaining."
- Average Linkage: Measures the average distance between all pairs of points in two clusters.

Advantages:

- Hierarchy Visualization: Hierarchical Clustering produces a dendrogram, allowing practitioners to explore the data's hierarchy and nested relationships.
- No Fixed Number of Clusters: Unlike K-Means, Hierarchical Clustering doesn't require specifying the number of clusters in advance.
- Robustness: Hierarchical Clustering's ability to consider all data points in the hierarchy can mitigate the effects of noise and outliers.

Limitations:

- Scalability: Hierarchical Clustering can become computationally expensive, especially for large datasets.
- Subjectivity: The choice of linkage criterion can impact results. Different criteria may lead to different cluster assignments.
- Dendrogram Interpretation: While dendrograms provide insights, determining the optimal number of clusters from them can be subjective.

Applications:

- Taxonomy and Biology: Hierarchical Clustering is used to classify species based on their genetic or phenotypic characteristics, creating taxonomic trees.
- Text Document Clustering: Hierarchical Clustering is employed to group similar documents into topics, enabling easier information retrieval.
- Market Segmentation: Hierarchical Clustering can assist in segmenting customers based on more complex criteria or behaviors.

- Image Segmentation: In computer vision, Hierarchical Clustering can segment images into regions based on color or texture similarities.
- Pharmacology: In drug discovery, Hierarchical Clustering helps group compounds with similar structures or properties.

Extensions and Variants:

- Agglomerative vs. Divisive: Agglomerative Hierarchical Clustering starts with individual data points and merges them into clusters. Divisive Hierarchical Clustering begins with all data points in one cluster and divides them into smaller clusters.
- Ward's Method: A linkage criterion that minimizes the variance within clusters during merging.
- Feature-Based Hierarchical Clustering: Instead of clustering data points, this variant clusters features or variables to identify groups of correlated features.

In conclusion, Hierarchical Clustering is a versatile technique that uncovers hierarchical relationships within data. Its ability to visualize hierarchical structures through dendrograms makes it valuable for understanding complex data relationships and creating taxonomies. Careful consideration of linkage criteria and the interpretability of dendrograms are essential for obtaining meaningful insights from Hierarchical Clustering.

11.3 Principal Component Analysis (PCA)

Principal Component Analysis (PCA) is a dimensionality reduction technique used to transform high-dimensional data into a lower-dimensional representation while retaining as much of the original variability as possible. It aims to find the directions (principal components) in the data space along which the data varies the most. PCA is widely used for visualization, noise reduction, and feature extraction. Let's explore the key concepts and applications of PCA:

Algorithm Overview:

- Data Centering: Center the data by subtracting the mean of each feature to ensure that the principal components represent variations around the origin.
- Covariance Matrix: Calculate the covariance matrix of the centered data to understand how features relate to each other.

- Eigenvalue Decomposition: Compute the eigenvalues and eigenvectors of the covariance matrix. The eigenvectors are the principal components.
- Selecting Principal Components: Sort the eigenvectors by their corresponding eigenvalues in decreasing order. Choose the top k eigenvectors to retain most of the variance.
- Projection: Project the centered data onto the selected principal components to obtain the lower-dimensional representation.

Advantages:

- Dimensionality Reduction: PCA reduces the number of features while preserving the most important information, making data more manageable and interpretable.
- Noise Reduction: High-dimensional data often contains noise. By focusing on the directions with the highest variance, PCA can suppress noise.
- Visualization: PCA allows data visualization in lower-dimensional space, facilitating insights and pattern recognition.

- Feature Extraction: PCA can transform data into a new set of features that may capture more relevant information for modeling.

Limitations:

- Linearity: PCA assumes linear relationships between features. Nonlinear variations may not be captured effectively.
- Information Loss: Reducing dimensionality inherently involves losing some information. The challenge lies in finding the right balance.
- Interpretability: The principal components are linear combinations of original features, which can make their interpretation less intuitive.

Applications:

- Image Compression: In computer vision, PCA is used to compress images by reducing the number of pixels while retaining essential features.

- Feature Engineering: PCA can transform raw features into a reduced set of features that capture most of the variance in the data.
- Data Visualization: By projecting data onto a lower-dimensional space, PCA enables visualization of high-dimensional data for better understanding.
- Data Preprocessing: PCA can be used as a preprocessing step to reduce the dimensionality of data before applying machine learning algorithms.
- Genetics: In genomics, PCA can be used to analyze gene expression data and identify patterns across different conditions.

Extensions and Variants:

- Kernel PCA: Extends PCA to capture nonlinear relationships by mapping data into a higher-dimensional space using kernel functions.
- Sparse PCA: Introduces sparsity to the principal components, resulting in a solution that uses only a subset of the original features.
- Incremental PCA: Suitable for processing large datasets, this method processes data in smaller batches to compute principal components incrementally.

[234]

In conclusion, Principal Component Analysis (PCA) is a powerful technique for reducing the dimensionality of high-dimensional data while preserving essential information. By identifying the most significant directions of variability, PCA enables more efficient data representation, visualization, and feature extraction. Careful consideration of its assumptions and application requirements is essential for utilizing PCA effectively in various domains of machine learning and data analysis.

11.4 t-Distributed Stochastic Neighbor Embedding (t-SNE)

t-Distributed Stochastic Neighbor Embedding (t-SNE) is a powerful dimensionality reduction technique primarily used for visualizing high-dimensional data in a lower-dimensional space. Unlike some other techniques, t-SNE focuses on preserving the relationships between data points, making it particularly effective for capturing intricate structures and clusters within the data. t-SNE is widely used in exploratory data analysis, clustering analysis, and visualization of complex datasets. Let's delve into the key concepts and applications of t-SNE:

Algorithm Overview:

- Pairwise Similarities: Calculate pairwise similarities between data points in the high-dimensional space. Gaussian distributions are often used to represent these similarities.
- Probability Distribution: Construct probability distributions that represent similarities between data

points in the high-dimensional space and their corresponding points in the lower-dimensional space.
- Perplexity: Perplexity is a hyperparameter that controls the balance between preserving local and global relationships. It determines the effective number of neighbors for each data point.
- Kullback-Leibler Divergence: The algorithm minimizes the Kullback-Leibler divergence between the two probability distributions, ensuring that similar points are modeled closely in the lower-dimensional space.
- Gradient Descent: Gradient descent is used to optimize the positions of data points in the lower-dimensional space to minimize the divergence.

Advantages:

- Preserving Local Structures: t-SNE excels at preserving local relationships and capturing clusters, making it suitable for visualizing complex data patterns.
- Nonlinearity: Unlike linear techniques like PCA, t-SNE can capture nonlinear relationships between data points.
- Cluster Identification: t-SNE can often reveal distinct clusters in the data, aiding in cluster analysis and understanding data groupings.

Limitations:

- Stochastic Nature: t-SNE's results can vary between runs due to its stochastic nature, making it important to run the algorithm multiple times for stability.
- Perplexity Selection: Choosing an appropriate perplexity value can be challenging, as it affects the balance between local and global structure preservation.
- Distances: t-SNE is sensitive to the choice of distance metric and similarity computation.

Applications:

- Visualization: t-SNE is widely used to visualize high-dimensional data in two or three dimensions, making complex data patterns more interpretable.
- Cluster Analysis: By revealing data clusters, t-SNE aids in identifying natural groupings within data.
- Dimensionality Reduction: While t-SNE is primarily a visualization technique, the lower-dimensional embeddings it produces can be used as features for subsequent analyses.

- Genomics: t-SNE is applied to visualize gene expression patterns and identify clusters of cells with similar genetic profiles.

Extensions and Variants:

- LargeVis: An extension of t-SNE designed for visualizing large datasets more efficiently.
- Barnes-Hut t-SNE: A faster variant of t-SNE that uses a Barnes-Hut tree for approximating pairwise similarities.

In conclusion, t-Distributed Stochastic Neighbor Embedding (t-SNE) is a powerful tool for visualizing complex high-dimensional data in a lower-dimensional space. Its ability to capture local relationships and reveal clusters makes it valuable for exploratory data analysis and understanding data structures. While it has limitations and requires careful parameter tuning, t-SNE remains an essential technique for gaining insights into data patterns that may not be apparent in the original high-dimensional space.

Chapter 12: Regularization and Regularized Regression

12.1 Ridge Regression

12.2 Lasso Regression

12.3 Elastic Net

12.4 Choosing Regularization Parameters

12.1 Ridge Regression

Ridge Regression, also known as Tikhonov regularization, is a technique used in statistics and machine learning to mitigate the problems of multicollinearity and overfitting in linear regression models. It's a form of regularized linear regression that adds a regularization term to the original linear regression cost function.

In standard linear regression, the goal is to find a set of coefficient values that minimize the sum of squared differences between the observed outcomes and the predictions made by the linear model. However, when there are correlated predictor variables (multicollinearity) or when the number of predictors is large relative to the number of observations, the model can become unstable and prone to overfitting.

Ridge Regression addresses these issues by introducing a penalty term that discourages the magnitude of the coefficient values from growing too large. This penalty term is directly related to the sum of the squared values of the coefficients. By adding this penalty to the original least squares cost function, Ridge Regression forces the optimization process to find coefficient values that are not only a good fit for the data but also relatively small in magnitude.

The strength of the regularization in Ridge Regression is controlled by a hyperparameter called the regularization parameter or "alpha." A larger alpha value leads to stronger regularization, causing the coefficients to be pushed closer to zero. On the other hand, a smaller alpha allows the coefficients to take on larger values, similar to what we see in standard linear regression.

Ridge Regression seeks to find the coefficient values that minimize the following cost function:

Cost = Sum of squared differences + alpha * Sum of squared coefficients

Here, the first term represents the standard least squares error, and the second term is the regularization penalty term.

[243]

The coefficient values are adjusted to strike a balance between fitting the data and keeping the coefficients small.

In summary, Ridge Regression is a valuable tool for dealing with multicollinearity and preventing overfitting in linear regression models. It achieves this by adding a regularization term to the cost function, which encourages the model to favor simpler coefficient values while still capturing the underlying relationships in the data.

12.2 Lasso Regression

Lasso Regression, short for "Least Absolute Shrinkage and Selection Operator," is another form of regularized linear regression that addresses multicollinearity and feature selection issues in predictive modeling. Like Ridge Regression, Lasso Regression adds a penalty term to the linear regression cost function to control the magnitude of coefficient values. However, Lasso Regression uses a slightly different penalty approach, which can lead to sparse coefficient solutions.

In standard linear regression, multicollinearity can lead to unstable and unreliable coefficient estimates, as correlated predictor variables can make it difficult to discern the individual contributions of each variable. Lasso Regression combats this issue by adding a penalty term proportional to the absolute values of the coefficients to the original cost function.

The main difference between Ridge and Lasso Regression lies in the penalty term. While Ridge uses the sum of squared coefficients, Lasso uses the sum of the absolute values of coefficients. This subtle change has a significant impact: Lasso has the ability to drive some coefficient values exactly to zero.

This feature inherently performs feature selection, effectively removing less relevant predictors from the model and producing a simpler model with fewer variables.

The cost function minimized by Lasso Regression is as follows:

Cost = Sum of squared differences + alpha * Sum of absolute coefficients

Similar to Ridge Regression, the regularization strength is controlled by the hyperparameter "alpha." Higher values of alpha result in stronger regularization, pushing more coefficients towards zero and favoring sparsity. Conversely, smaller alpha values allow coefficients to take on larger values.

Lasso Regression's ability to perform feature selection makes it particularly useful when dealing with high-dimensional datasets with many predictors, as it can automatically identify and focus on the most important variables. However, it's important to note that Lasso may struggle with correlated predictors since it tends to select only one predictor out of a group of correlated ones, while the others are pushed to zero.

In summary, Lasso Regression is a regularization technique that combines the benefits of linear regression and feature selection. By introducing an absolute value penalty term to the cost function, Lasso promotes sparse coefficient solutions, effectively excluding less relevant predictors and providing a simpler and more interpretable model.

12.3 Elastic Net

Elastic Net is a hybrid regularization technique that combines the characteristics of both Ridge Regression and Lasso Regression. It aims to address the limitations of each method by incorporating both the L2 (ridge) and L1 (lasso) penalty terms in the linear regression cost function. By doing so, Elastic Net seeks to leverage the strengths of both regularization techniques while mitigating their individual weaknesses.

In cases where the dataset has high multicollinearity and a large number of predictors, Lasso Regression might struggle to perform well due to its tendency to select only a subset of predictors and ignore the others. Additionally, Ridge Regression may still leave many coefficients relatively large. Elastic Net aims to overcome these limitations by introducing a new parameter, "l1_ratio," which controls the balance between the L1 and L2 penalties.

The Elastic Net cost function includes both the sum of squared differences (as in standard linear regression), the sum of squared coefficients (as in Ridge Regression), and the sum of absolute coefficients (as in Lasso Regression). The cost function is as follows:

Cost = Sum of squared differences + alpha * ((1 - l1_ratio) * Sum of squared coefficients + l1_ratio * Sum of absolute coefficients)

Here, "alpha" controls the overall strength of regularization, similar to Ridge and Lasso Regression. The "l1_ratio" parameter determines the balance between the L1 and L2 penalties. When "l1_ratio" is set to 1, Elastic Net is essentially equivalent to Lasso Regression. When "l1_ratio" is set to 0, it becomes equivalent to Ridge Regression. For values between 0 and 1, Elastic Net blends the properties of both methods.

Elastic Net is particularly useful when the dataset has many features and some of them are highly correlated. It allows for automatic feature selection like Lasso, but by incorporating Ridge's penalty term, it can also handle cases where multiple correlated features should be retained together. This balance makes Elastic Net more versatile and robust in situations where Lasso or Ridge alone might not perform optimally.

[249]

In summary, Elastic Net is a regularization technique that combines Ridge and Lasso Regression by introducing a parameter that balances between their respective penalties. This balance provides a flexible approach for handling multicollinearity, feature selection, and coefficient magnitude control in linear regression models.

12.4 Choosing Regularization Parameters

Choosing the appropriate regularization parameters, such as "alpha" for Ridge and Lasso Regression or "alpha" and "l1_ratio" for Elastic Net, is a crucial step in applying these regularization techniques effectively. These parameters control the strength of regularization and the balance between different penalty terms, influencing the model's complexity, bias-variance trade-off, and generalization performance. Here's how you can approach choosing regularization parameters:

Grid Search or Cross-Validation: One common approach is to perform a grid search over a range of potential parameter values. You can create a grid of alpha values (and l1_ratio for Elastic Net) and evaluate the model's performance using cross-validation. Cross-validation helps assess how well the model generalizes to new, unseen data by splitting the dataset into training and validation subsets multiple times.

Cross-Validation Scores: During grid search or cross-validation, monitor the performance metrics, such as mean squared error (MSE), root mean squared error (RMSE), or mean absolute error (MAE), on the validation sets for different parameter combinations. The goal is to find the parameter values that result in the best trade-off between bias and variance.

Regularization Path: For Ridge and Lasso Regression, it can be helpful to plot the regularization path, which shows how the coefficients change as the regularization parameter varies. This can provide insights into which coefficients are being shrunk to zero and how quickly the model is regularizing.

L-Curve Method: For Ridge Regression, you can use the L-curve method. It involves plotting the cross-validation error against the L2 norm of the coefficients. The point where the error starts to level off while the norm of the coefficients is still relatively small can indicate a good choice of alpha.

Information Criteria: Some information criteria, like Akaike Information Criterion (AIC) and Bayesian Information Criterion (BIC), can help in selecting regularization parameters. These criteria balance model fit with model complexity and can guide you toward a simpler model that still captures the essential patterns in the data.

Domain Knowledge: Sometimes, domain knowledge can provide insights into reasonable ranges for regularization parameters. For example, if you know that most predictors are likely to be relevant, you might lean towards lower alpha values.

Nested Cross-Validation: To get a more accurate estimate of the model's generalization performance, you can use nested cross-validation. In this approach, an outer loop performs model evaluation using cross-validation, while an inner loop performs the grid search for selecting the best parameters. This helps prevent overfitting on the validation sets used in parameter selection.

Regularization Path Stability: For Elastic Net, you can also assess the stability of the selected features across different splits of the data during cross-validation. Features that consistently appear across multiple splits are more likely to be robust and reliable predictors.

[253]

Remember that the optimal parameter values can vary based on the specific dataset and the problem you're addressing. It's important to strike a balance between model complexity and generalization performance. Regularization parameters should be chosen based on rigorous evaluation techniques rather than relying solely on intuition.

[254]

Chapter 13: Decision Trees and Ensemble Learning

13.1 Decision Tree Construction

13.2 Random Forests

13.3 Gradient Boosting

13.4 XGBoost and LightGBM

13.1 Decision Tree Construction

Decision Trees are a popular machine learning algorithm used for both classification and regression tasks. They work by recursively partitioning the input feature space into subsets that are as homogeneous as possible with respect to the target variable. The construction of a decision tree involves selecting features to split on and determining the splitting criteria at each node. Here's an overview of the decision tree construction process:

Root Node: The initial step is to select the best feature to split the dataset. The chosen feature is the one that best separates the data based on some criterion, often aiming to minimize impurity or maximize information gain.

Splitting Criteria: The chosen splitting criterion depends on the type of task. For classification, common criteria include Gini impurity and entropy, which measure the level of disorder or randomness in a set of labels. For regression, mean squared error (MSE) is a common criterion that quantifies the variance of target values within subsets.

Splitting Process: Once the initial feature is selected, the dataset is split into subsets based on the possible values of that feature. Each subset corresponds to a branch stemming from the root node. The process then repeats recursively for each subset.

Stopping Criteria: The recursion continues until a stopping criterion is met. Stopping criteria could include reaching a maximum depth of the tree, having a minimum number of samples in a node, or reaching a threshold in impurity reduction.

Leaf Nodes: When the stopping criteria are met, terminal nodes, known as leaf nodes, are created. These nodes represent the final predictions for classification or regression tasks. For classification, the majority class in the leaf node is often chosen as the predicted class. For regression, the average or median target value of the samples in the leaf node is used as the prediction.

Pruning (Optional): After constructing the full tree, pruning can be applied to prevent overfitting. Pruning involves removing branches that do not contribute significantly to improving model performance on validation data. This leads to a simpler and more generalizable tree.

Decision Trees are easy to visualize and interpret, making them valuable for understanding the decision-making process of a model. However, they are prone to overfitting, especially when they become deep and complex. This is where ensemble methods, such as Random Forests and Gradient Boosting, come into play.

Ensemble methods combine multiple decision trees to create more robust and accurate models. They leverage the diversity of individual decision trees and combine their predictions in a way that reduces overfitting and improves overall performance.

[259]

13.2 Random Forests

Random Forests is an ensemble learning method that improves the performance of decision trees by combining the predictions of multiple individual trees. It reduces overfitting and increases predictive accuracy through a process called bagging (bootstrap aggregating) and feature randomization. Here's an overview of how Random Forests work:

Bagging: Bagging involves creating multiple decision trees using random subsets of the training data. Each tree is trained on a different subset of the data obtained through random sampling with replacement. This introduces diversity among the trees and helps to reduce overfitting.

Feature Randomization: In addition to using random subsets of the data, Random Forests also perform feature randomization. For each tree and at each split, a random subset of features is considered for splitting the node. This randomness further increases the diversity among the trees and improves the generalization ability of the ensemble.

Decision Tree Construction: Each decision tree in the Random Forest is constructed as described in the previous section. However, due to the use of random subsets of data and feature subsets, the individual trees are typically deeper and more complex.

Voting (Classification) or Averaging (Regression): When making predictions, each tree in the Random Forest produces its own prediction. For classification tasks, the final prediction is determined by majority voting among the individual trees' predictions. For regression tasks, the final prediction is the average of the predictions from all the trees.

Advantages: Random Forests are robust against overfitting, thanks to the combination of bagging and feature randomization. They tend to generalize well to new, unseen data. They can handle high-dimensional data and are less sensitive to outliers and noisy data points compared to individual decision trees.

Tuning: Random Forests have hyperparameters that can be tuned, such as the number of trees in the ensemble and the maximum depth of individual trees. Cross-validation or other validation techniques can help in selecting optimal hyperparameters.

Feature Importance: Random Forests provide a measure of feature importance, which indicates the contribution of each feature to the model's predictions. This information can be useful for feature selection and understanding the underlying patterns in the data.

Random Forests are widely used for various tasks, including classification, regression, and feature selection. They strike a balance between complexity, interpretability, and performance, making them a powerful tool in machine learning.

13.3 Gradient Boosting

Gradient Boosting is another ensemble learning method that builds an additive model in a forward stepwise manner, combining the predictions of multiple weak learners (typically decision trees) to create a strong predictive model. Unlike Random Forests, which focus on reducing variance through bagging and feature randomization, Gradient Boosting focuses on reducing bias and improving predictive accuracy through iterative optimization. Here's how Gradient Boosting works:

Initialization: The process starts with an initial model, often a simple one like the mean value for regression tasks or a uniform class distribution for classification tasks.

Residual Calculation: For each data point in the training set, the difference between the actual target value and the predicted value from the current model is calculated. These differences are called residuals.

Building Weak Learners: A weak learner, typically a decision tree with a shallow depth (a "weak" tree), is trained to predict these residuals. The weak learner focuses on capturing the patterns in the residuals that the current model couldn't explain.

Step Size (Learning Rate): A small fraction of the predictions from the weak learner (residual predictions) is added to the current model's predictions. The magnitude of this fraction is determined by a hyperparameter called the learning rate. A lower learning rate can make the learning process more robust but slower.

Iterative Process: Steps 2 to 4 are repeated iteratively. In each iteration, a new weak learner is trained to predict the residuals of the previous model. The predictions from all the weak learners are combined, and the residuals are updated. The process continues until a predefined number of iterations is reached or until a certain level of performance is achieved.

Combining Weak Learners: The final prediction from the Gradient Boosting model is the sum of the predictions from all the weak learners, each scaled by the learning rate. This additive nature of the model allows it to gradually correct errors made by previous models.

Regularization (Optional): Gradient Boosting models can also include regularization techniques to prevent overfitting. One common approach is to add a penalty term to the loss function that measures the complexity of the model.

Hyperparameter Tuning: Gradient Boosting has hyperparameters such as the number of weak learners, the learning rate, and the maximum depth of the weak learners. These parameters need to be tuned using techniques like grid search or random search.

Gradient Boosting is a powerful technique that often outperforms other machine learning algorithms due to its ability to capture complex relationships in the data. It's widely used in both regression and classification tasks, and popular implementations include XGBoost, LightGBM, and CatBoost.

13.4 XGBoost and LightGBM

XGBoost is an optimized and highly efficient implementation of the Gradient Boosting algorithm. It was designed to improve upon the limitations of traditional Gradient Boosting and has become one of the most popular and powerful machine learning libraries. XGBoost is known for its robustness, scalability, and high performance. Here are some key features and characteristics of XGBoost:

Regularization: XGBoost incorporates both L1 (Lasso) and L2 (Ridge) regularization terms into its cost function. This helps prevent overfitting and contributes to better generalization.

Handling Missing Values: XGBoost can handle missing values during the training and prediction processes. It learns the optimal direction to handle missing data points.

Built-in Cross-Validation: XGBoost supports built-in cross-validation, which simplifies the process of finding the optimal number of boosting rounds.

Tree Pruning: It applies a depth-first approach during tree construction and prunes trees using "max_depth" parameter to control their complexity.

Parallel Processing: XGBoost supports parallel processing, making it faster and more efficient, especially for large datasets.

Flexible Objective Functions: It allows users to define their own custom objective functions, which is useful for specific problem domains.

Weighted Instances: XGBoost can assign different weights to different instances, which is helpful when dealing with imbalanced datasets.

LightGBM (Light Gradient Boosting Machine):

LightGBM is another popular gradient boosting framework that is specifically designed for speed and efficiency. It's developed by Microsoft and is known for its ability to handle large datasets and outperform other gradient boosting implementations in terms of training time.

Here are some key features and characteristics of LightGBM:

Histogram-Based Approach: LightGBM uses a histogram-based approach to bin continuous features, which significantly reduces memory usage and speeds up the training process.

Gradient-Based One-Side Sampling: LightGBM uses a gradient-based approach for selecting data samples for tree construction, leading to faster convergence and reduced overfitting.

Leaf-Wise Tree Growth: Unlike traditional depth-first tree growth, LightGBM uses a leaf-wise tree growth strategy, which can lead to deeper trees and better model performance.

Categorical Feature Support: LightGBM can handle categorical features directly, without the need for one-hot encoding.

Parallel and GPU Learning: LightGBM supports parallel processing and GPU acceleration, further enhancing its speed and scalability.

Monotonic Constraints: It allows users to define monotonic constraints for features, which enforces a desired relationship between features and the target variable.

Both XGBoost and LightGBM are highly efficient, and the choice between them often depends on the specific dataset, problem complexity, and available resources. They have a wide range of hyperparameters that can be tuned to achieve the best performance on a given task.

Chapter 14: Neural Network Architectures

14.1 Autoencoders

14.2 Generative Adversarial Networks (GANs)

14.3 Transformers

14.4 Applications in Generation and NLP

14.1 Autoencoders

Autoencoders are a type of artificial neural network architecture primarily used for unsupervised learning tasks, such as dimensionality reduction, feature learning, and data compression. They are designed to encode input data into a lower-dimensional representation and then decode it back to its original form, aiming to learn a compressed representation that captures essential features of the data. Autoencoders consist of an encoder and a decoder, which work together to learn an efficient data representation.

Here's an overview of the components and functioning of autoencoders:

Encoder: The encoder takes the input data and maps it to a lower-dimensional representation, often referred to as the "encoding" or "latent" space. The encoding is a compressed representation of the input data that captures its most important features.

Decoder: The decoder takes the encoded representation from the encoder and attempts to reconstruct the original input data from it. The goal is to minimize the reconstruction error, encouraging the network to learn a meaningful encoding.

Loss Function: The reconstruction error is used as the loss function during training. The difference between the input data and the reconstructed output is minimized through backpropagation, adjusting the weights of the neural network's layers.

Bottleneck Layer: The layer in the middle of the network, corresponding to the encoding, is often referred to as the "bottleneck" layer. It has a lower dimensionality than the input and output layers, and it forces the network to capture the most relevant features of the data.

Autoencoders can have various architectures, each with its own characteristics and use cases:

Standard Autoencoder: The simplest form of autoencoder, consisting of an encoder, a decoder, and a bottleneck layer. It's primarily used for dimensionality reduction and feature learning.

Sparse Autoencoder: Introduces sparsity constraints to the encoding, encouraging some neurons in the bottleneck layer to be inactive. This can result in more robust and informative representations.

Denoising Autoencoder: Trained to reconstruct clean data from noisy inputs. It helps the network learn to extract useful features and patterns from corrupted data.

Variational Autoencoder (VAE): A probabilistic version of autoencoders that learns to generate data in the encoding space. VAEs are used for generating new data samples and for performing tasks like image synthesis.

Convolutional Autoencoder: Utilizes convolutional layers for image data, allowing it to capture spatial hierarchies and patterns present in images.

Recurrent Autoencoder: Incorporates recurrent layers to handle sequential data like time series or text.

Autoencoders find applications in various domains, including image denoising, anomaly detection, image compression, and feature extraction. They are also a fundamental component of more advanced techniques like generative adversarial networks (GANs) and transfer learning.

14.2 Generative Adversarial Networks (GANs)

Generative Adversarial Networks, commonly known as GANs, are a type of neural network architecture introduced by Ian Goodfellow and his colleagues in 2014. GANs are designed for generating new data samples that resemble a given training dataset. The idea behind GANs is to have two neural networks, the generator and the discriminator, competing against each other in a game-like fashion.

Here's how GANs work and the roles of their components:

Generator: The generator's task is to create new data samples that resemble the training data. It takes random noise as input and produces synthetic data samples. Initially, the generator produces random and meaningless outputs.

Discriminator: The discriminator, also known as the critic, aims to distinguish between real data from the training set and fake data produced by the generator. It takes both real and synthetic data as input and assigns a probability that the input data is real.

The training process of GANs involves a back-and-forth competition between the generator and the discriminator:

Training the Discriminator: The discriminator is trained on real data samples from the training set and fake data samples generated by the generator. It learns to assign high probabilities to real data and low probabilities to fake data.

Training the Generator: The generator is trained to produce data samples that can fool the discriminator. It attempts to improve its output quality by generating samples that the discriminator struggles to distinguish from real data.

Adversarial Process: The training process alternates between training the discriminator and training the generator. The goal is to reach a point where the generator creates data that is indistinguishable from real data.

Loss Functions: GANs use specific loss functions for the generator and the discriminator. The generator aims to minimize the discriminator's ability to correctly classify generated data, while the discriminator aims to correctly classify real and fake data.

Convergence: Ideally, as training progresses, the generator becomes more skilled at generating realistic data, and the discriminator becomes more challenged in distinguishing real from fake data. In the optimal scenario, the generator produces data that is practically indistinguishable from real data.

GANs have led to remarkable breakthroughs in various domains, including image generation, style transfer, data augmentation, and even text-to-image synthesis. However, training GANs can be challenging, as maintaining the balance between the generator and discriminator is delicate, and the training process can be prone to mode collapse (when the generator produces a limited range of samples). Despite these challenges, GANs have revolutionized the field of generative modeling and have inspired numerous variations and applications.

14.3 Transformers

Transformers are a groundbreaking neural network architecture introduced by Vaswani et al. in the paper "Attention Is All You Need" in 2017. They were initially designed for natural language processing tasks, such as machine translation, but have since become the foundation for many state-of-the-art models in various domains, including language, vision, and even protein folding. Transformers are particularly effective for tasks that involve sequences of data, making them versatile and powerful.

The key innovation in Transformers is the self-attention mechanism, which enables the model to weigh the importance of different words or elements in a sequence when making predictions. This mechanism allows Transformers to capture long-range dependencies and relationships in data more effectively than traditional recurrent or convolutional networks.

Here's an overview of the components and functioning of Transformers:

Self-Attention: The self-attention mechanism computes a weighted sum of the input sequence elements, where the weights are determined by the relationships between elements. It assigns higher weights to elements that are more relevant for each prediction.

Multi-Head Attention: To capture different aspects of relationships, Transformers use multiple attention mechanisms in parallel, each referred to as a "head." This multi-head attention mechanism enhances the model's ability to focus on various patterns and relationships.

Positional Encoding: Unlike recurrent networks that inherently capture sequence order, Transformers lack this property. To overcome this, positional encodings are added to the input embeddings to provide information about the position of each element in the sequence.

Encoder and Decoder Stacks: Transformers consist of an encoder and a decoder stack. The encoder processes the input sequence and generates a representation, while the decoder generates the output sequence. Both encoder and decoder contain multiple layers of self-attention and feed-forward neural networks.

Attention Masks: For tasks that involve sequence-to-sequence mapping, such as machine translation, attention masks are used to ensure that predictions are made based only on the available information at each step.

Positional-wise Feed-Forward Networks: After the self-attention mechanism, each position in the sequence undergoes a feed-forward neural network transformation that captures complex interactions between elements.

Layer Normalization and Residual Connections: Similar to other deep neural networks, Transformers use layer normalization and residual connections to stabilize and speed up training.

Transformers have paved the way for various models, including BERT, GPT, T5, and more, each designed to excel in specific tasks. They've revolutionized natural language processing tasks, enabling models to understand context, contextually generate text, perform document classification, and much more. Additionally, their architecture has been adapted to computer vision tasks, such as image captioning and object detection, further highlighting their versatility and effectiveness across different domains.

14.4 Applications in Generation and NLP

Neural network architectures have been instrumental in various generation tasks, where the goal is to generate new data samples that are coherent, realistic, and similar to the training data. Some notable applications include:

Image Generation: Generative models like Generative Adversarial Networks (GANs) and Variational Autoencoders (VAEs) have been used to create realistic images from random noise. They have found applications in art generation, image-to-image translation, and data augmentation.

Text Generation: Recurrent neural networks (RNNs), Transformers, and specifically language models like GPT-3 are used for text generation. This includes tasks like chatbots, story generation, code completion, and content creation.

Music Generation: RNNs and variations like LSTM (Long Short-Term Memory) networks have been employed for generating music sequences. These models can compose melodies and harmonies in various musical styles.

Video Generation: Temporal generative models combine CNNs (Convolutional Neural Networks) and RNNs to generate videos or video frames. These models have applications in video prediction, animation, and video synthesis.

Applications in Natural Language Processing (NLP):

NLP involves the use of neural network architectures to understand, interpret, and generate human language. Neural networks have driven significant advancements in NLP tasks:

Machine Translation: Sequence-to-sequence models, often based on Transformers, have revolutionized machine translation by producing more fluent and accurate translations between different languages.

Sentiment Analysis: Recurrent and convolutional neural networks are used for sentiment analysis, where the sentiment (positive, negative, neutral) of a text is determined.

Named Entity Recognition (NER): NER systems employ neural networks to identify entities like names of people, places, and organizations within a text.

Text Classification: Neural networks are widely used for classifying text into predefined categories, such as news categorization, spam detection, and topic modeling.

Question Answering: Models like BERT and its variants have achieved state-of-the-art performance in question-answering tasks, where machines answer questions based on given contexts.

Language Generation: Neural networks can generate coherent and contextually relevant language, such as automatic text summarization, dialogue systems, and content generation.

Language Translation: Neural machine translation models, especially those built upon Transformer architectures, have greatly improved the quality of language translation systems.

Speech Recognition: Recurrent and convolutional neural networks are used for converting spoken language into text, enabling applications like voice assistants and transcription services.

Language Understanding: Embeddings and contextual representations from neural networks are used to understand nuances in language, enabling improved language understanding and semantic understanding.

These applications demonstrate how neural network architectures have reshaped generation tasks and revolutionized natural language processing, allowing machines to understand, interpret, and generate human language with unprecedented accuracy and quality.

Chapter 15: Future Trends in Machine Learning

15.1 Explainable AI

15.2 Federated Learning

15.3 Quantum Machine Learning

15.4 Ethical Considerations

15.1 Explainable AI

Explainable AI (XAI) refers to the concept of designing and developing machine learning models and algorithms that can provide understandable explanations for their decisions and predictions. As machine learning models become more complex and sophisticated, there's a growing need to understand why they make certain decisions, especially in critical applications like healthcare, finance, and autonomous systems. XAI aims to bridge the gap between the "black box" nature of advanced models and the human need for transparency and interpretability.

Key aspects of Explainable AI include:

Interpretability: XAI focuses on making machine learning models more interpretable by humans. This involves creating models that not only produce accurate predictions but also provide insights into how those predictions are derived.

User Trust: Explainable models increase user trust in AI systems. Users are more likely to accept and adopt AI technologies if they can understand and verify the reasoning behind the system's decisions.

Regulatory Compliance: In domains like healthcare and finance, regulatory bodies often require explanations for AI-based decisions. XAI can help organizations comply with regulations by providing transparent explanations.

Bias and Fairness: XAI can help detect and mitigate biases in AI models. When the decision-making process is transparent, it becomes easier to identify and rectify bias and discrimination issues.

Model Debugging: Explainable AI tools can assist in diagnosing issues and errors in machine learning models, making it easier to identify the source of unexpected behavior.

Human-AI Collaboration: XAI promotes collaboration between humans and AI systems. When humans understand how AI systems arrive at conclusions, they can provide valuable insights and domain expertise.

Different Techniques: Various techniques contribute to XAI, such as feature importance analysis, local explanations (explaining individual predictions), global explanations (explaining model behavior as a whole), and model visualization.

Algorithmic Transparency: Some XAI methods focus on making existing complex models more transparent, while others involve developing inherently interpretable models.

Trade-off with Performance: There's often a trade-off between model complexity and performance versus interpretability. Highly interpretable models might sacrifice some predictive accuracy.

Human-Centered Design: XAI involves considering the needs and expectations of users during the model development process, ensuring that the explanations provided are meaningful and relevant to the end-users.

Explainable AI is especially important in high-stakes applications where understanding the reasoning behind AI decisions is critical. As AI technologies continue to evolve, the demand for transparency and accountability will likely

[290]

increase. Researchers and practitioners are actively working on developing and improving techniques for creating explainable models that strike a balance between performance and interpretability.

15.2 Federated Learning

Federated Learning is a distributed machine learning approach that enables model training across multiple devices or servers while keeping the data localized and decentralized. In traditional machine learning, data is collected from various sources, centralized, and used to train a model. Federated Learning, on the other hand, brings the model to the data rather than the data to the model. This has important implications for privacy, security, and efficiency, especially in scenarios where data cannot be easily shared or centralized.

Key concepts and characteristics of Federated Learning include:

Decentralized Training: In Federated Learning, training occurs on the devices or servers where the data resides. The model is sent to these locations, and each location performs local training using its data.

Privacy Preservation: Federated Learning addresses privacy concerns by avoiding the need to share raw data. Only model updates, which are aggregated, are shared, protecting sensitive information.

Data Ownership: Data owners retain control over their data, reducing the need to centralize and share sensitive or proprietary information.

Bandwidth Efficiency: Federated Learning reduces the need for transmitting large datasets to a central server, saving bandwidth and reducing network communication costs.

Local Model Updates: On-device training produces local model updates that reflect the characteristics of each specific dataset. These updates are then aggregated to improve the global model.

Aggregation and Averaging: The local model updates are aggregated, often using techniques like weighted averaging or gradient-based methods, to update the global model.

Heterogeneous Data: Federated Learning supports learning from heterogeneous data sources, such as devices with varying distributions, types, and sizes of data.

Robustness and Edge Computing: Federated Learning can improve model robustness by training on diverse data sources. It's also relevant for edge computing scenarios, where devices process data locally.

Challenges: Federated Learning comes with challenges like communication efficiency, model synchronization, and dealing with data distribution heterogeneity.

Applications: Federated Learning is valuable in scenarios like healthcare (patient data privacy), IoT devices (resource-constrained devices), and organizations with data privacy restrictions.

Security and Encryption: Federated Learning requires secure communication and encryption methods to prevent unauthorized access to model updates.

Federated Learning aligns well with emerging trends in privacy-preserving machine learning and the increasing awareness of data privacy issues. It allows organizations to harness the benefits of machine learning while respecting data ownership and privacy concerns. As the technology continues to develop, it has the potential to unlock new opportunities for collaborative model training across distributed environments.

15.3 Quantum Machine Learning

Quantum Machine Learning (QML) is an emerging interdisciplinary field that combines principles from quantum computing and machine learning. It explores how quantum computers can potentially enhance various aspects of machine learning tasks, such as optimization, data analysis, and pattern recognition. Quantum computers leverage the principles of quantum mechanics, which can provide advantages over classical computers for certain types of problems.

Key concepts and characteristics of Quantum Machine Learning include:

Quantum Advantage: Quantum computers can potentially solve certain problems more efficiently than classical computers by exploiting phenomena like superposition and entanglement.

Quantum Bits (Qubits): Quantum computers use qubits instead of classical bits. Qubits can exist in a superposition of states, which allows quantum computers to process multiple possibilities simultaneously.

Quantum Entanglement: Entanglement is a unique quantum property where qubits become interconnected, even when separated. This property can be leveraged to process information in novel ways.

Quantum Gates: Quantum computers operate using quantum gates that manipulate qubits' states. These gates perform complex operations that classical logic gates cannot achieve.

Quantum Circuits: Quantum algorithms are represented as quantum circuits, analogous to classical algorithms represented as flowcharts. These circuits describe the sequence of quantum gates applied to qubits.

Quantum Algorithms: Quantum computers have the potential to accelerate certain algorithms, like factorization, optimization, and database search, which have implications for machine learning tasks.

Quantum Annealing: Quantum annealers are specialized quantum devices designed to solve optimization problems by minimizing energy functions. They're applicable to some machine learning optimization tasks.

Quantum Neural Networks: Quantum versions of neural networks, called quantum neural networks or quantum circuits, are being explored for tasks like function approximation and data classification.

Hybrid Approaches: Due to the current limitations of quantum computers, hybrid approaches that combine classical and quantum computing methods are often used.

Challenges: QML faces challenges such as qubit noise, limited qubit coherence, and hardware constraints. Error correction techniques are crucial for scaling up quantum computations.

Applications: QML holds potential for tasks like solving complex optimization problems, simulating quantum systems, and improving machine learning algorithms.

[298]

Quantum Machine Learning is an area of ongoing research and development, with significant challenges to overcome before it reaches its full potential. While practical, large-scale quantum computers are not yet widely available, QML has already shown promising results for specific problems. As quantum technology advances, it has the potential to revolutionize various aspects of machine learning and computational science.

15.4 Ethical Considerations

As machine learning technologies become more pervasive and impactful, ethical considerations have become crucial to ensuring that these technologies are developed, deployed, and used responsibly. Ethical considerations in machine learning encompass a range of concerns related to fairness, transparency, accountability, bias, privacy, and the potential societal impact of AI systems. Addressing these concerns is essential to build trustworthy and beneficial AI systems.

Key ethical considerations in machine learning include:

Fairness and Bias: Machine learning models can inherit biases present in the training data, leading to unfair or discriminatory outcomes. Ensuring fairness and addressing bias is crucial to prevent harm to underrepresented groups and promote equitable outcomes.

Transparency and Explainability: Black-box models that produce results without explanations can erode user trust. Efforts to make models transparent and provide explanations for decisions are vital for accountability and understanding.

[300]

Accountability: As AI systems make important decisions, it's crucial to attribute responsibility and accountability for both positive and negative outcomes. Ensuring clear lines of accountability helps prevent harm and facilitates learning from mistakes.

Privacy: Machine learning systems often handle sensitive personal data. Protecting user privacy through data anonymization, encryption, and robust data security practices is essential to maintain user trust.

Data Collection and Consent: Ethical data collection practices involve obtaining informed consent from individuals whose data is being used. Transparent communication about data usage and providing the option to opt out are important considerations.

Algorithmic Transparency: Understanding how algorithms work, including the criteria they use for decision-making, is important for users and stakeholders. Lack of algorithmic transparency can lead to suspicion and distrust.

Benefit and Harm: It's crucial to assess the potential benefits and risks of AI systems. Ensuring that the benefits significantly outweigh the potential harm is a key ethical principle.

Accounting for Social Impact: Developers and organizations should consider the broader societal impact of AI technologies. This includes addressing economic disparities, job displacement, and other social implications.

Unintended Consequences: AI systems can have unintended consequences that were not anticipated during development. Thorough testing, monitoring, and addressing potential negative outcomes are essential.

Regulation and Policies: Governments and organizations are developing regulations and policies to guide the responsible development and deployment of AI technologies. These frameworks aim to ensure ethical standards are met.

Ethical AI Education: Training data scientists, engineers, and AI practitioners in ethical considerations is essential to ensure that ethical concerns are embedded in the design and development process.

[302]

Ethical considerations in machine learning are dynamic and evolve as technology advances. The AI community, researchers, practitioners, policymakers, and the public are actively engaged in discussions to define ethical standards and guidelines for the responsible use of AI. Balancing innovation and ethical considerations is key to building AI systems that benefit society as a whole while minimizing potential negative impacts.

Appendix: Mathematical Notation, Concepts, examples, exercises and solutions

- Common Symbols and Notations
- Review of Key Mathematical Concepts
- Exercises and Solution (Brief explanations)

Review of Key Mathematical Concepts

Note: Below are brief overviews of key mathematical concepts.

Algebra:

- **Concept:** Quadratic Equations
- **Explanation:** Quadratic equations have the form $ax^2 + bx + c = 0$, where a, b, and c are constants.
- **Example:** Solve the quadratic equation $2x^2 - 5x + 3 = 0$.

Calculus:

- **Concept:** Derivative of a Function

- **Explanation:** The derivative of a function $f(x)$ at a point measures its rate of change.
- **Example:** Find the derivative of $f(x) = 3x^2 - 2x + 1$.

Linear Algebra:

- **Concept:** Matrix Multiplication
- **Explanation:** Multiplying two matrices involves combining rows and columns to calculate new values.
- **Example:** Compute the product of matrices $\mathbf{A}$ and $\mathbf{B}$.

Exercises and Solutions (A Brief overview)

Chapter 1: Foundations of Linear Algebra

- **Exercise:** Calculate the determinant of the matrix $\mathbf{A} = \begin{bmatrix} 3 & -1 \\ 2 & 4 \end{bmatrix}$.
- **Solution:** The determinant of $\mathbf{A}$ is calculated as $\text{det}(\mathbf{A}) = (3 \cdot 4) - (-1 \cdot 2) = 14$.

Chapter 2: Multivariable Calculus

- **Exercise:** Find the gradient of the function $f(x, y) = 2x^2 + 3y^3$.
- **Solution:** The gradient is given by $\nabla f = \begin{bmatrix} \frac{\partial f}{\partial x} \\ \frac{\partial f}{\partial y} \end{bmatrix} = \begin{bmatrix} 4x \\ 9y^2 \end{bmatrix}$.

Chapter 3: Probability and Statistics

- **Exercise:** Given two independent random variables X and Y with variances $\text{Var}(X) = 4$ and $\text{Var}(Y) = 9$, find the variance of $2X - 3Y$.
- **Solution:** The variance of the linear combination is $\text{Var}(2X - 3Y) = 2^2 \cdot \text{Var}(X) + (-3)^2 \cdot \text{Var}(Y) = 4 \cdot 4 + 9 \cdot 9 = 97$.

Chapter 4: Information Theory

- **Exercise:** Compute the entropy of a fair coin toss.
- **Solution:** The entropy of a fair coin toss is $H(X) = -\left(\frac{1}{2} \log_2\frac{1}{2} + \frac{1}{2} \log_2\frac{1}{2}\right) = 1$ bit.

Chapter 5: Linear Regression

- **Exercise:** Perform multiple linear regression on a dataset with two features, x_1 and x_2, and a target variable y. Interpret the coefficients.

- **Solution:** The multiple linear regression equation is $y = \beta_0 + \beta_1 x_1 + \beta_2 x_2$. The coefficients β_1 and β_2 represent the change in the target variable for a unit change in x_1 and x_2 while holding other variables constant.

Chapter 6: Classification

- **Exercise:** Given a binary classification problem with true positive (TP) = 30, false positive (FP) = 10, true negative (TN) = 50, and false negative (FN) = 20, calculate the accuracy and F1-score.
- **Solution:** Accuracy = $\frac{TP + TN}{TP + TN + FP + FN} = \frac{30 + 50}{30 + 50 + 10 + 20} = 0.8$. F1-score = $\frac{2 \cdot \text{Precision} \cdot \text{Recall}}{\text{Precision} + \text{Recall}}$.

Chapter 7: Support Vector Machines

- **Exercise:** Given a linearly separable dataset, explain the concept of the margin in a support vector machine.

- **Solution:** The margin is the distance between the decision boundary and the nearest data point from either class. A larger margin indicates better generalization to unseen data.

Chapter 8: Neural Networks and Deep Learning Basics

- **Exercise:** Describe the purpose of activation functions in neural networks.
- **Solution:** Activation functions introduce non-linearity, allowing neural networks to learn complex relationships between features and improve their capacity to model a wide range of functions.

Chapter 9: Convolutional Neural Networks (CNNs)

- **Exercise:** Explain the concept of pooling layers in CNNs and their role in reducing spatial dimensions.
- **Solution:** Pooling layers downsample the spatial dimensions of the input while retaining important features. Common pooling types include max pooling and average pooling.

Chapter 10: Recurrent Neural Networks (RNNs)

- **Exercise:** Define the vanishing gradient problem in recurrent neural networks.
- **Solution:** The vanishing gradient problem occurs when gradients in backpropagation become extremely small in deep RNNs, leading to slow convergence and difficulty in learning long-range dependencies.

Chapter 11: Unsupervised Learning: Clustering and Dimensionality Reduction

- **Exercise:** Describe the K-Means clustering algorithm and its steps.
- **Solution:** K-Means is an iterative algorithm that partitions data into 'k' clusters. It starts with initial centroids, assigns points to the nearest centroid, recalculates centroids, and repeats until convergence.

Chapter 12: Regularization and Regularized Regression

- **Exercise:** Compare Ridge and Lasso regularization techniques.

- **Solution:** Ridge adds L2 regularization by penalizing large coefficient values, while Lasso adds L1 regularization and can lead to sparse coefficients, effectively performing feature selection.

Chapter 13: Decision Trees and Ensemble Learning

- **Exercise:** Explain the concept of bagging in ensemble methods.
- **Solution:** Bagging (Bootstrap Aggregating) involves training multiple models on random subsets of the training data with replacement and averaging their predictions to reduce overfitting.

Chapter 14: Neural Network Architectures

- **Exercise:** Describe the purpose of autoencoders in neural network architectures.
- **Solution:** Autoencoders are used for unsupervised learning and dimensionality reduction by learning efficient data representations in an encoder-decoder structure.

Chapter 15: Future Trends in Machine Learning

- **Exercise:** Describe the concept of federated learning and its advantages.
- **Solution:** Federated learning enables model training on decentralized data while keeping data locally, improving privacy and data security.

[314]

Mathematics for Machine Learning: A Deep Dive into Algorithms

Nibedita Sahu

About the Author

Nibedita is a passionate tech enthusiast with a background in Mathematics. With a keen interest in both mathematics and machine learning, Nibedita brings a unique perspective to the intricate intersection of these fields. Through this book, she shares her expertise and insights to empower readers with a strong foundation in the mathematical principles that underpin modern machine learning algorithms.

Unlock the Power of Mathematics in Machine Learning

In "Mathematics for Machine Learning: A Deep Dive into Algorithms," Nibedita guides readers on an enlightening journey through the intricate world of mathematical concepts that drive machine learning innovation. From linear algebra to information theory, from calculus to probability, this book equips you with the tools to understand the "why" behind machine learning techniques and to apply them effectively.

Hands-On Learning

With practical examples, explanations and exercises with solutions, you'll not only grasp abstract mathematical concepts but also witness their real-world applications. Through clear explanations and insightful insights, you'll build a solid foundation that empowers you to explore the full potential of machine learning algorithms.

Your Path to Mastery

Whether you're a student, developer, or data scientist, "Mathematics for Machine Learning" offers a roadmap to deepen your understanding and harness the power of mathematics in the realm of artificial intelligence. Embark on a journey that transforms equations into insights and intelligence into innovation.

Prepare to embark on a captivating exploration of mathematics in machine learning.

Get your copy of "Mathematics for Machine Learning: A Deep Dive into Algorithms" and elevate your understanding of the mathematical foundations that shape the future of AI.

Happy exploring!